OT 34
Operator Theory: Advances and Applications
Vol. 34

Birkhäuser Verlag
Basel · Boston · Berlin

Orthogonal Matrix-valued Polynomials and Applications

Seminar on Operator Theory at the School of Mathematical Sciences, Tel Aviv University

Edited by

I. Gohberg

1988

Birkhäuser Verlag
Basel · Boston · Berlin

Volume Editorial Office:

School of Mathematical Sciences
Raymond and Beverly Sackler Faculty of Exact Sciences
Tel Aviv University
Ramat Aviv
Israel

Library of Congress Cataloging in Publication Data

Orthogonal matrix valued polynomials and applications : seminar on
operator theory at the School of Mathematical Sciences, Tel Aviv
University / edited by I. Gohberg.
p. cm. -- (Operator theory, advances and applications ; vol. 34)
Includes bibliographies.
ISBN 0-8176-2242-X (U.S.)
1. Orthogonal polynomials -- Congresses. 2. Matrices -- Congresses.
I. Gohberg, I. (Israel), 1928- . II. Series: Operator theory,
advances and applications ; v. 34.
QA404.5.073 1988 88-22307
515'.55 -- dc 19

CIP-Titelaufnahme der Deutschen Bibliothek

Orthogonal matrix valued polynomials and applications /
Seminar on Operator Theory at the School of Mathemat.
Sciences, Tel Aviv Univ. Ed. by I. Gohberg. – Basel ; Boston ;
Berlin : Birkhäuser, 1988
(Operator theory ; 34)
ISBN 3-7643-2242-X (Basel ...) Pb.
ISBN 0-8176-2242-X (Boston) Pb.
NE: Gochberg, Izrail' [Hrsg.]; Seminar on Operator Theory <1987 –
1988, Rāmat-Āvîv>; Bêt has-Sēfer le-Maddā'ê ham-Mātēmātîqā
<Rāmat-Āvîv>; Krejn, Mark G.: Festschrift; GT

Printed in Germany
ISBN 3-7643-2242-X
ISBN 0-8176-2242-X

Dedicated to
M. G. Krein
on the Occasion of
his Eightieth Birthday

Table of Contents

EDITORIAL INTRODUCTION

This volume consists of six papers which further develop the theory of orthogonal matrix valued polynomials which are orthogonal on the unit circle with respect to an indefinite matrix valued weight. The distribution of zeros of such polynomials is a central theme which is in fact touched upon by each of the papers. It is thus more than fitting that this volume be dedicated to M.G. Krein who initiated the study of this problem for scalar polynomials more than twenty years ago.

This volume contains generalizations of the above mentioned results to nonstationary analogues of orthogonal polynomials. Connections with the theory of reproducing kernel spaces with indefinite inner product and assorted interpolation problems are also discussed.

During the spring semester of 1987 and 1988 the meetings of the seminar on operator theory at the School of Mathematics of the Tel Aviv University were mostly devoted to matrix valued orthogonal polynomials. Almost all of the papers in this volume are extended versions of talks given at these seminars.

The authors of this volume dedicate it to M.G. Krein on the occasion of his eightieth birthday (with an apology for the delay). It is an expression of their respect and admiration. The volume also contains a complete bibliography of M.G. Krein until 1988.

BIBLIOGRAPHY OF

MARK GRIGOR'EVICH KREIN *

[1] Le systèm dérivé et les contours dérivés, Zh. Nauchn. Issled Kafedr. Odessa, 2. No. 3 (1926), 1-13.

[2] On the Taylor series defining an analytic function which is regular in a domain bounded by several circles, Izv. Fiz.-Mat. O.-va, Kazan' (3) **2** (1927), 50-57.

[3] Über den Satz von "Curvatura Integra", Izv.Fiz.-Mat. O.-va, Kazan' (3) **3** (1928), 36-47.

[4] L'intègrale de Stieltjes dans la théorie des contours convexes, Izv. Fiz. Mat. O.-va, Kazan' (3) **3** (1929), 81-93.

[5] Sur l'aire mixte de deux ovales, Zap. Prir.-Tekn. Otd. Akad. Nauk Ukr. SSR (1929), 3.

[6] Zur Strukturfrage von Orthogonalmatrizen, Zap. Prir.-Tekhn. Otd. Akad. Nauk Ukr. SSR (1929), 1-8 (with (F.R. Gantmakher).

[7] On normal operators in Hermitian space, Izv. Fiz.-Mat. O.-va, Kazan' (3) **4** (1929-1930), 71-184 (with F.R. Gantmakher).

[8] Ergänzungen zu der Abhandlung "Zur Strukturfrage von Orthogonalmatrizen", Zap. Prir.-Tekhn. Otd. Akad. Nauk Ukr. SSR **1** (1931), 103-108.

[9] A contribution to the theory of symmetric polynomials, Mat. Sb., **40** (1933), 271-283.

[10] On the theory of vibrations of multisupported beams. Vestnik Inzh. i Tekhn., **4** (1933).

[11] On the spectrum of a Jacobi form in connection with the theory of torsional oscillations of shafts. Mat. Sb., **40** (1933), 455-466.

---

* Translated by Andrei Iacob

[12] Über eine neue Klasse von Hermiteschen Formen und über eine Verallgemeinerung des trigonometrischen Momenten problems, Izv. Akad. Nauk SSSR Otd. Mat. Estestven. Nauk **9** (1933), 1259-1275.

[13] Über eine Transformation der Bezoutiante, die Sturmschen Satze fuhrt, Zap. Khar'kov. Mat. O.-va. (4) **10** (1933), 33-40 (with M.A. Naimark).

[14] On the nodes of harmonic oscillations of mechanical systems of a special form, Mat. Sb., **41** (1934), 339-348.

[15] Über Fouriersche Reihen beschränkter summierbarer Funktionen und ein neues Extremumproblem. I. Soobshch, Nauchno-Issled. Inst. Mat. Mekh. Khar'kov Gos. Univ. i. Khar'kov. Mat. O.-va (4) **9** (1934), 9-23 (with N.I. Akhiezer).

[16] Über Fouriersche Reihen beschrankter Summierbarer Funktionen und ein neues Extremumproblem. II, Soobshch. Nauchno-Issled. Inst. Mat. Mekh. Khar'kov Gos. Univ. i. Khar'kov. Mat. O.-va (4) **10** (1934), 3-32 (with N.I. Akhiezer).

[17] On the Fourier series of bounded summable functions, Proc. 2nd All-Union Math. Conf., vol. 2, Leningrad (1934), 151 (with N.I. Akhiezer).

[18] On a generalization of the investigations of Academician A.A. Markov on limiting values of integrals, Proc. 2nd All-Union Math. Conf., Vol.2, Leningrad (1934), 152-154.

[19] On integral equations with oscillating fundamental functions, Proc. 2nd All-Union Math. Conf., vol. 2, Leningrad (1934), 259-262.

[20] A general method of forming the frequency equations of vibrating plane frames. Trudy Inst. Inzh. Vodn. Transp., Odessa 1 (1934) (with Ya. L. Nudel'man).

[21] On integral kernels of Green-function type, Tr. Univ.-a Odessa **1** (1935), 39-50 (with F.R. Gantmakher).

[22] On applications of the Bezoutian to problems of localization of roots of algebraic. Odessa, Tr. Univ. **1** (1935), 51-69 (with M.A. Naimark), (Ukrainian).

[23] On a special class of differential operators, Dokl. Akad. Nauk SSSR **2** (1935), 345-349.

[24] Sur les dérivées de noyaux de Mercer, C.R. Acad. Sci., Paris, **200** (1935), 797-799.

[25] Sur une formule de quadrature de Tchebycheff, C.R. Acad. Sci. Paris **200** (1935), 890-893 (with N.I. Akhiezer).

[26] Sur les equations intégrales chargées, C.R. Acad. Sci. Paris, **201** (1935), 24-26.

[27] Sur quelques applications des noyaux de Kellog aux problemes d'oscillation, Soobshch. Nauchn.-Issled. Inst. Mat. Mekh. Khar'kov. Univ. i, Khar'kov Mat. O.va (4) **11** (1935), 3-20.

[28] Über eine Transformation der reellen Toeplitzschen Formen und das Momentproblem in einem endlichen Intervalle, Soobshch. Nauchn.-Issled. Inst. Mat. Mekh. Khar'kov. Gos. Univ. i Khar'kov. Mat. O.-va (4) **11** (1935), 21-26 (with N.I. Akhiezer).

[29] On a special class of determinants in connection with Kellog's integral kernels, Mat. Sb., **42** (1935), 501-508 (with F.R. Gantmakher).

[30] Sur les matrices oscillatoires, C.R. Acad. Sci. Paris, **201** (1935), 577-579 (with F.R. Gantmakher).

[31] The method of symmetric and Hermitian forms in the theory of separation of the roots of algebraic equations, Khar'kov (1936), 1-41 (with M.A. Naimark); English transl.: Linear and Multilinear Algebra **10** (1981), 265-308.

[32] On Green functions positive in the sense of Mercer, Dokl. Akad. Nauk SSSR **1** (1936), 55-58.

[33] On bilinear decompositions of symmetric kernels positive in the sense of Mercer, Dokl. Akad. Nauk SSSR **1** (1936), 303-306 (with A.M. Danilevskii).

[34] On two minimum problems connected with the moment problem, Dokl. Akad. Nauk SSSR **1** (1936), 331-334 (with N.I. Akhiezer).

[35] On oscillation differential operators, Dokl. Akad. Nauk SSSR **4** (1936), 379-382.

[36] Sur les vibrations propres des tiges dont l'une des extremites est encastree et l'autre libre, Zap. Khar'kov. Mat. O.-va (4) **12** (1936), 3-11.

[37] Das Momentenproblem bei der zusätzlichen Bedingung von A. Markoff, Soobshch. Nauchn.-Issled. Inst. Mat. Mekh. Khar'kov. Gos. Univ. i Khar'kov. Mat. O.-va (4) **12** (1936), 13-36 (with N.I. Akhiezer).

[38] Bemerkung zur Arbeit "Über Fouriersche Reihen beschränkter summierbarer Funktionen und ein neues Extremumproblem", Soobshch. Nauch.-Issled. Inst. Mat. Mekh. Khar'kov. Gos. Univ. i Khar'kov. Mat.O.-va (4) **12** (1936), 37-40 (with N.I. Akhiezer).

[39] Sur quelques propriétés de noyaux de Kellog, Zap. Khar'kov. Mat. O.-va (4) **13** (1936), 15-28.

[40] On P.F. Pankovich's paper "On a form of the basic differential equations of small oscillations of a system without gyroscopic terms", Prikl. Mat. Mekh. **1** (1936), 159-161 (with F.R. Gantmakher).

[41] On the characteristic numbers of symmetric differentiable kernels, Mat. Sb., **2** (**44**) (1937), 725-732.

[42] Sur les développements des fonctions arbitraires en séries de fonctions fondamentales d'un probleme aux limites quelconque, Mat. Sb., **2** (**44**) (1937), 923-934.

[43] Sur les operateurs différentiels autoadjoints et leurs fonctions de Green symmétriques, Mat. Sb., **2** (**44**) (1937) 1032-1072.

[44] The moment problem on two intervals with the additional conditions of A.A. Markov, Zap. Khar'kov. Mat. O.-va (4), **14**, (1937) 47-60 (with N.I. Akhiezer).

[45] On some properties of the Kellog's resolvent kernel. Zap. Khar'kov Mat. O.-VA (4) **14** (1937), 61-74.

[46] On positive additive functionals on linear normed spaces, Zap. Khar'kov. Mat. O.-va (4) **14** (1937), 227-237

[47] On several questions concerning the geometry of convex sets belonging to a linear, normed, and complete space, Dokl. Akad. Nauk SSSR **14** (1937), 5-8.

[48] On the best approximation of differentiable periodic functions by trigonometric sums, Dokl. Akad. Nauk SSSR **15** (1937), 107-112 (with N.I. Akhiezer).

[49] Sur les matrices complètement non-négatives et oscillatoires, Comp. Math., **4** (1937), 445-476 (with F.R. Gantmakher).

[50] Concerning the theory of vibrations of rod systems, Trudy Inst. Mat., Odessa 12 (1938), 193-225 (with Ya. L. Nudel'man).

[51] Some questions in the Theory of Moments, Gos. Nauchn.-

Tekhn. Izd.-vo Ukr., Khar'kov (1938), (with N.I. Akhiezer); English Transl.: Transl. Math. Monographs, Vol. 2, Amer. Math. Soc., Providence, R.I. (1962).

[52] Some remarks on the coefficients of quadrature formulae of Gaussian type, Trudy Odess'kogo Derzh. Univ. Mat. 2 (1938), 29-38 (with N.I. Akhiezer).

[53] On the Nevanlinna-Pick problem, Trudy Odess. Gos. Univ. Mat., 2 (1938), 63-77 (with P.G. Rekhtman).

[54] On minimaximal properties of nodes of obertones of a vibrating rod. Odessa, Trudy Univ., Ser. Mat. 2 (1938), 103-112 (with Ya. L. Nudel'man) (Ukrainian).

[55] On the theory of best approximation of periodic functions Dokl. Akad. Nauk SSSR **18** (1938), 245-250).

[56] On the best approximation of continuous differentiable functions on the full real axis, Dokl. Akad. Nauk SSSR **18** (1938), 619-624.

[57] On the linear operators that leave invariant a conic set, Dokl. Akad. Nauk SSSR **23** (1939), 749-752.

[58] On totally nonnegative Green functions of ordinary differential operators, Dokl. Akad. Nauk SSSR **24** (1939), 220-223 (with G.M. Finkel'shtein).

[59] On nonsymmetric oscillating Green functions of ordinary differential operators, Dokl. Akad. Nauk SSSR **25** (1939), 643-646.

[60] Oscillation theorems for ordinary linear differential operators of arbitrary order, Dokl. Akad. Nauk SSSR **25** (1939), 717-720.

[61] On the decomposition of a linear functional into positive components, Dokl. Akad. Nauk SSSR **25** (1939), 721-724 (with Yu. I. Grosberg).

[62] On some quadrature formulae of P.L. Chebyshev and A.A. Markov, Memorial volume dedicated to D.A. Grave, Moscow (1940), pp. 15-28 (with N.I. Akhiezer).

[63] On weighted integral equations whose distribution functions are not monotonic, Memorial volume dedicated to D.A. Grave, Moscow (1940), pp. 88-103.

[64] On a property of bases in Banach space, Zap. Khar'kov. Mat. O.-va (4) **16** (1940), 106-110 (with D.P. Mil'man and M.A. Rutman).

[65] Some remarks about three papers of M.S. Verblunsky, Zap. Nauchn.-Issled. Inst. Mat. Mekh. i Khar'kov. Mat.O.-va (4) **16** (1940), 129-134 (with N.I. Akhiezer).

[66] On some minimum-problems in the class of Stepanoff almost periodic functions. Soobsch. Nauchn.-Issled. Inst. Mat. Mekh. Khar'kov. Gos. Univ. i Khar'kov. Mat. O.-va (4) **17** (1940), 111-124 (with B.M. Levitan)

[67] On extreme points of regularly convex sets, Studia Math., **9** (1940), 133-138 (with D.P. Mil'man).

[68] On regularly convex sets in the space conjugate to a Banach space, Ann. Math., **41** (1940), 556-583 (with V.L. Shmul'yan).

[69] On the continuation problem for Hermitian-positive continuous functions, Dokl. Akad. Nauk SSSR **26** (1940),71-21.

[70] On an intrinsic characteristic of the space of all continuous functions defined on a bicompact Hausdorff space, Dokl. Akad. Nauk SSSR **27** (1940), 427-431 (with S.G. Krein).

[71] Basic properties of normal conic sets in Banach space, Dokl. Akad. Nauk SSSR **28** (1940), 13-17.

[72] On the minimal decomposition of a linear functional into positive components, Dokl. Akad. Nauk SSSR **28** (1940), 18-22.

[73] On a ring of functions defined on a topological group, Dokl. Akad. Nauk SSSR **29** (1940), 275-280.

[74] On a special ring of functions, Dokl. Akad. Nauk SSSR **29** (1940), 355-359.

[75] On a new property of the Sturm-Liouville operator, Trudy Univ., Odessa, Ser. Mat., 3 (1940), 21-32.

[76] On the theory of almost-periodic functions on a topological group, Dokl. Akad. Nauk SSSR **30** (1941), 5-8.

[77] On positive functionals on almost-periodic functions, Dokl. Akad. Nauk SSSR **30** (1941), 9-12.

[78] On a generalization of Plancherel's theorem to the case of Fourier integrals on a commutative topological group, Dokl. Akad. Nauk SSSR **30** (1941), 482-486.

[79] Sur l'espace des fonctions continues définies sur un bicompact de Hausdorff et ses sousespaces semi-ordonnés, Mat. Sb., **13** (55) (1943), 1-38 (with S.G. Krein).

[80] On the representation of functions by Fourier-Stieltjes integrals, Uchen. Zap. Kuibyshev. Gos. Pedag. i Uchit. Inst., **7** (1943), 123-148.

[81] On Hermitian operators with deficiency indices equal to one, I, Dokl. Akad. Nauk SSSR **43** (1944), 339-342.

[82] On Hermitian operators with deficiency indices equal to one. II, Dokl. Akad. Nauk SSSR **44** (1944), 143-146.

[83] On a remarkable class of Hermitian operators, Dokl. Akad. Nauk SSSR **44** (1944), 191-195.

[84] On a generalized moment problem, Dokl. Akad. Nauk SSSR **44** (1944), 239-243.

[85] On the logarithm of an infinitely decomposable Hermitian-positive function, Dokl. Akad. Nauk SSSR **45** (1944), 99-102.

[86] On the continuation problem for helical arcs in Hilbert space, Dokl. Akad. Nauk SSSR **45** (1944), 147-150.

[87] On a generalization of investigations of G. Szego, V.I. Smirnov, and A.N. Kolmogorov, Dokl. Akad. Nauk SSSR **46** (1945) 95-98.

[88] On an extrapolation problem of A.N. Kolmogorov, Dokl. Akad. Nauk SSSR **46** (1945), 339-342.

[89] On self-adjoint extensions of bounded and semibounded Hermitian operators, Dokl. Akad. Nauk SSSR **48** (1945), 323-326.

[90] On the resolvents of a Hermitian operator with deficiency index (m,m), Dokl. Akad. Nauk SSSR **52** (1946), 657-660.

[91] On a general method of decomposing positive-definite kernels into elementary products, Dokl. Akad. Nauk SSSR **53** (1946), 3-6.

[92] On a theorem of M. Ya. Vygodstkii, Mat. Sb., **18 (60)** (1946), 447-450.

[93] The basic theorems concerning the extension of Hermitian operators and some of their applications to the theory of orthogonal polynomials and the moment problem, Usp. Mat. Nauk **2**, no.3 (1947), 60-106 (with M.A. Krasnosel'skii).

[94] On a general method of decomposing positive-definite kernels into elementary ones, Usp. Mat. Nauk **2**, no.3 (1947), p. 181.

[95] The theory of self-adjoint extensions of semibounded Hermitian operators and its applications. I, Mat. Sb., **20** (**62**) (1947), 431-498.

[96] The theory of self-adjoint extensions of semibounded Hermitian operators and its applications. II, Mat. Sb., **21** (**63**) (1947), 365-404.

[97] On linear complete continuous operators in functional spaces with two norms., Zh. Inst. Mat. Akad. Nauk. SSR **9** (1947), 365-404 (Ukrainian).

[98] On the theory of entire functions of exponential type, Izv. Akad. Nauk SSSR Ser. Mat., **11** (1947), 309-326.

[99] Linear operators leaving invariant a cone in Banach space, Usp. Mat. Nauk **3**, no. 1 (1948), 3-95 (with M.A. Rutman); English transl.: Amer. Math. Soc. Transl. (1), **10** (1962), 199-325.

[100] Helical lines in an infinite-dimensional Lobachevsky space and Lorentz transformations, Usp. Mat. Nauk 3, no.3 (1948), 158-160.

[101] On some problems connected with Lyapunov's circle of ideas in stability theory, Usp. Mat. Nauk **3**, no. 3 (1948) 166-169.

[102] On Hermitian operators with directing functions, Sb. Trudov Inst. Mat. Akad. Nauk Ukr. SSR 10 (1948), 83-106.

[103] On deficiency numbers of linear operators in Banach space and on certain geometric questions, Sb. Trudov Mat. Inst. Akad. Nauk Ukr. SSR 11 (1948), 97-112 (with M.A. Krasnosel'skii and D.P. Mil'man).

[104] Functional Analysis, in :"**Mathematics in the USSR over 30 years**", Moscow, Leningrad (1948), 608-672 (with L.A. Lyusternik).

[105] Fundamental aspects of the representation theory of Hermitian operators with deficiency index (m,m), Ukr. Mat. Zh., **1**, no.2 (1949), 3-66; English transl.: Amer. Math. Soc. Transl. (2) **97** (1970), 75-143.

[106] Hermitian-positive kernels on homogeneous spaces, I, Ukr. Mat. Zh., **1**, no.4 (1949), 64-98; English transl.: Amer. Math. Soc. Transl. (2) **34** (1963), 69-108.

[107] Infinite J-matrices and the matrix moment problem, Dokl. Akad. Nauk SSSR **69** (1949), 125-128.

[108] On entire almost-periodic functions of exponential type,

Dokl. Akad. Nauk SSSR **69** (1949), 285-287 (with B. Ya. Levin).

[109] The duality principle for a bicompact group and square block-algebra, Dokl. Akad. Nauk SSSR **69** (1949), 726-728.

[110] Hermitian-positive kernels on homogeneous spaces, II, Ukr. Mat. Zh., **2**, no. 1 (1950), 10-59; English transl.: Amer. Math. Soc. Transl. (2) **34** (1963), 109-164.

[111] On an application of the fixed point principle in the theory of linear transformations of spaces with an indefinite metric, Usp. Mat. Nauk **5**, no.2 (1950), 180-190; English transl.: Amer. Math. Soc. Transl. (2) 1 (1955), 27-35.

[112] Remark on a possible generalization of a theorem of A. Haar and A.N. Kolmogorov, Usp. Mat. Nauk **5**, no.1 (1950), 217-229 (with S.I. Zukhovitskii).

[113] Generalization of certain investigations of A.M.Lyapunov on linear differential equations with periodic coefficients, Dokl. Akad. Nauk **73** (1950), 445-448.

[114] On the Sturm-Liouville boundary value problem on the interval $(0,\infty)$ and on a class of integral equations, Dokl. Akad. Nauk SSSR **73** (1950), 1125-1128.

[115] On a one-dimensional singular boundary value problem of even order on the interval $(0,\infty)$, Dokl. Akad. Nauk SSSR **74** (1950), 9-12.

[116] On some investigations of A.M. Liapunov concerning differential equations with periodic coefficients, Dokl. Akad. Nauk SSSR **75** (1950), 495-498 (with K.P.Kovalenko).

[117] Oscillation Matrices, Kernels, and Small Oscillations of Mechanical Systems, 2nd edition, Moscow, Leningrad (1950) (with F.R. Gantmakher); German transl.: Akademic Verlag, Berlin (1960).

[118] Concerning the mathematical theory of the method of amplitude-phase modulation, Sb. Trudov Inst. Elektrotekhn. Akad. Nauk SSSR **7** (1951), 96-104 (with B.I. Korenblyum and S.I. Tetel'baum).

[119] On the theory of entire matrix-functions of exponential type, Ukr. Mat. Zh., **3**, no. 2 (1951).

[120] Solution of the inverse Sturm-Liouville problem, Dokl. Akad. Nauk SSSR **76** (1951), 21-24.

[121] Determination of the density of a nonhomogeneous symmetric string from its frequency spectrum, Dokl. Akad. Nauk SSSR **76** (1951), 345-348.

[122] On the application of an algebraic proposition in the theory of monodromy matrices, Usp. Mat. Nauk **6**, no.1 (1951), 171-177.

[123] On certain problems of the minimum and maximum of characteristic values and on the Lyapunov zones of stability, Prikl. Mat. Mekh., **15** (1951), 323-348; English transl.: Amer. Math. Soc. Transl. (2) **1** (2955), 163-187.

[124] The ideas of P.L. Chebyshev and A.A. Markov in the theory of limited values of integrals and their further development, Usp. Mat. Nauk **6**, no.4 (1951), 3-120 (with the redactional participation of P.G. Rekhtman); Amer. Math. Soc. Transl. (2) **12** (1959), 1-121.

[125] On a generalization of the Schwarz and Loewner lemmas, Zap. Mat. Otd. Fiz.-Mat. Fak. i Khar'kov. Mat. O.-va (4) **23** (1952), 95-101 (with N.I. Akhiezer).

[126] Stability of the index of an unbounded operator, Mat. Sb., **30** (**72**) (1952), 219-224 (with M.A. Krasnosel'skii).

[127] On an indetermined case of the Sturm-Liouville boundary value problem on the interval $(0,\infty)$, Izv. Akad. Nauk SSSR Ser. Mat., **16** (1952), 293-324.

[128] On inverse problems for a nonhomogeneous string, Dokl. Akad. Nauk SSSR **82** (1952), 669-672.

[129] On a generalization of investigations of Stieltjes, Dokl. Akad. Nauk SSSR **87** (1952), 881-884.

[130] On some new problems of the oscillation theory for Sturmian systems, Prikl. Mat. Mekh., **14** (1952), 555-568.

[131] On a trace formula in perturbation theory, Mat. Sb., **33** (**75**) (1953), 597-626.

[132] On the transfer function of a one-dimensional second-order boundary value problem, Dokl. Akad. Nauk SSSR **88** (1953), 405-408.

[133] Analogue of the Chebyshev-Markov inequalities in a one-dimensional boundary-value problem, Dokl. Akad. Nauk SSSR **89** (1953), 5-8.

[134] On some cases of effective determination of the density of a nonhomogeneous string from its spectral function, Dokl. Akad. Nauk SSSR **93** (1953), 617-620.

[135] On inverse problems of the theory of filters and λ-zones of stability. Dokl. Akad. Nauk **93** SSSR (1953), 767-770.

[136] A fundamental approximation problem in the theory of extrapolation and filtration of stationary stochastic processes, Dokl. Akad. Nauk SSSR **94** (1954), 13-16; English transl.: Selected Transl. in Math. Statistics and Probability, Inst. of Math. Statistics and Amer. Math. Soc. **4** (1963), 127-131.

[137] On a method of effective resolution of an inverse boundary value problem, Dokl. Akad. Nauk SSSR **94** (1954), 987-990.

[138] On integral equations that generate second-order differential equations, Dokl. Akad. Nauk SSSR **97** (1954), 21-24.

[139] The theory of spectral and transfer functions of one-dimensional boundary value problems and its applications, Usp. Mat. Nauk **9**, no. 3 (1954), p. 221.

[140] Development in a new direction of the Chebyshev-Markov theory of limiting values of integrals, Usp. Mat. Nauk **10** no. 1 (1955), 67-78 (with P.G. Rekhtman); English transl.: Amer. Math. Soc. Transl. (2) **12** (1959), 123-135.

[141] On a new method of solving linear integral equations of first and second kind, Dokl. Adkad. Nauk SSSR **100** (1955), 413-416.

[142] On the determination of the potential of a particle from its S-function, Dokl. Akad. Nauk SSSR **105** (1955), 433-436.

[143] Continual analogues of propositions on polynomials orthogonal on the unit circle, Dokl. Akad. Nauk SSSR **105** (1955), 637-640.

[144] On tests for stable boundedness of solutions of periodic canonical systems, Prikl. Mat. Mekh., **19** (1955), 641-680; English transl.: Amer. Math. Soc. Transl. (2) **120** (1983) 71-110.

[145] The basic propositions of the theory of λ-zones of stability of a canonical system of linear differential equations with periodic coefficients, Memorial volume dedicated to A.A. Andronov, Moscow (1955), 413-498; English trans.: Topics in Differential and Integral Equations and Operator Theory, Operator Theory: Advances and Applications, Vol. 7, Birkhäuser, Basel (1983), pp. 1-105, (1983), 1-70.

[146] On some basic propositions of the theory of systems of integral equations on the half-line with kernels depending on the difference of the arguments, Proc. 3rd All-Union Math. Conf., Vol. 2 (1956), pp. 37-38 (with I.C. Gohberg).

[147] Spectral theory of operators in spaces with an indefinite metric, Usp. Mat. Nauk **11**, no.4 (1956), 169-171 (with I.S. Yokhvidov).

[148] Spectral theory of operators in spaces with an indefinite metric. I. Tr. Mosk. Mat. O.-va **5** (1956), 367-432 (with I.S. Yokhvidov); English trans.: Amer. Math. Soc. Transl. (2) **13** (1960), 105-175.

[149] On the theory of accelerants and S-matrices of canonical differential systems, Dokl. Akad. Nauk SSSR **111** (1956), 1167-1170.

[150] Remark to the paper "Spectral theory of operators in spaces with indefinite metric. I," Tr. Mosk. Mat. O.va **6** (1957), p. 486 (with I.S. Yokhvidov).

[151] The basic propositions on defect numbers, root numbers, and indices of linear operators, Usp. Mat. Nauk **12**, no.2 (1957), 43-118 (with I.C. Gohberg); English trans.: Amer. Math. Soc. Transl. (2) **13** (1960), 185-264.

[152] On Bari bases in Hilbert space, Usp. Mat. Nauk **12**, no.3 (1957), 333-341.

[153] On the characteristic function $A(\lambda)$ of a linear canonical system of second-order differential equations with periodic coefficients, Prikl. Mat. Mekh., **21** (1957) 320-329.

[154] On a continual analogue of a formula of Christoffel from the theory of orthogonal polynomials, Dokl. Akad. Nauk SSSR **113** (1957), 970-973.

[155] Integral equations on the half-line with kernel depending on the difference of the arguments, Usp. Mat. Nauk **13**, no. 5 (1958), 3-120; English transl.: Amer. Math. Soc. Transl. (2) **22** (1962), 163-288.

[156] Criterion for the discreteness of the spectrum of a singular string, Izv. Vyssh. Uchebn. Zaved., Mat., 2 (1958), 136-153 (with I.S. Kats).

[157] On a pair integral equation and its transpose. I, Teor. Prikl. Mat., 1 (1958), 58-81 (with I.C. Gohberg);

English trans.: Topics in Differential and Integral Equations and Operator Theory, Operator Theory: Advances and Appl., vol. 7, Birkhäuser, Basel (1983), pp. 243-274.

[158] On the stability of the system of partial indices of the Hilbert problem with several unknown functions, Dokl. Akad. Nauk SSSR **119**, (1958), 854-857 (with I.C.Gohberg).

[159] Systems of integral equations on the half-line with kernels depending on the differences of arguments, Usp. Mat. Nauk **13**, no.2 (1958), 3-72 (with I.C. Gohberg); English trans.: Amer. Math. Soc. Transl. (2) **14** (1960), 217-287.

[160] On completely continuous operators with a spectrum concentrated at zero, Dokl. Akad. Nauk SSSR **128** (1959), 227-230 (with I.C. Gohberg).

[161] Criteria for completeness of the system of root vectors of a dissipative operator, Usp. Mat. Nauk **14**, no.3 (1959), 145-152; English transl.: Amer. Math. Soc. Transl. (2) **26** (1963), 221-229.

[162] On an integral representation of a continuous Hermitian-indefinite function with a finite number of negative squares, Dokl. Akad. Nauk SSSR **125** (1959), 31-34.

[163] Spectral theory of operators in spaces with an indefinite metric. II, Tr. Mosk. Mat. O.-va **8** (1959), 413-496 (with I.S. Yokhvidov); English transl.: Amer. Math. Soc. Transl. (2) **34** (1963), 283-373.

[164] On the theory of linear non-self-adjoint operators, Dokl. Akad. Nauk SSSR **130** (1960), 254-256; English transl.: Soviet Math. Dokl., 1 (1960), 38-40.

[165] Hamiltonian systems of linear differential equations with periodic coefficients, Kiev (1961), 1-55 (with V.A. Yakubovich).

[166] On the theory of transmission bands of periodic waveguides, Prikl. Mat. Mekh., **25** (1961), 24-37 (with G. Ya. Lyubarskii); English transl.: J. Appl. Math. Mech., **25** (1961), 29-48.

[167] On the theory of triangular representations of non-self-adjoint operators, Dokl. Akad. Nauk SSSR **137** (1961), 1034-1037 (with I.C. Gohberg); English transl.: Soviet Math. Dokl. **2** (1961), 392-395.

[168] Transformations of kernels of integral equations and their influence on the spectra of these equations, Ukr. Mat. Zh., **13**, no.3 (1961), 12-38 (with I.C. Gohberg); English transl.: Amer. Math. Soc. Transl. (2) **35** (1964) 263-295.

[169] On Volterra operators with imaginary component in one class or another, Dokl. Akad. Nauk SSSR **139** (1961), 779-782 (with I.C. Gohberg); English transl.: Soviet Math. Dokl. **2** (1961), 983-986.

[170] An extremum problem for polynomials, Ann. Univ. Sci. Budapest Eötvös Sect. Math. 3-4 (1961), 9-14 (with N.I. Akhiezer).

[171] On the theory of wave operators and scattering operators, Dokl. Akad. Nauk SSSR **144** (1962), 475-478 (with M. Sh. Birman); English transl.: Soviet Math. Dokl. **3** (1962), 740-744.

[172] On the problem of factoring operators in a Hilbert space, Dokl. Akad. Nauk SSSR **147** (1962), 279-282 (with I. C. Gohberg); English transl.: Soviet Math. Dokl. **3** (1962), 1578-1582.

[173] Analytic properties of multipliers of periodic canonical differential systems of positive type, Izv. Akad. Nauk SSSR Ser. Mat., **26** (1962), 549-572 (with G. Ya. Lyubarskii): English transl.: Amer. Math. Soc. Transl. (2) **89** (1970), 1-20.

[174] On perturbation determinants and a trace formula for unitary and self-adjoint operators, Dokl. Akad. Nauk SSSR **144** (1962), 268-271; English Transl.: Soviet Math. Dokl. **3** (1962), 707-710.

[175] Hamiltonian systems of linear differential equations with periodic coefficients, Proc. Internat. Symp. Nonlinear Vibrations, Kiev, Vol. 1 (1963), pp.277-305 (with V.A. Yakubovich); English transl.: Amer. Mat. Soc. Transl. (2) **120** (1983), 139-168.

[176] Some topics of the theory of wave and scattering operators, Joint Symposium Soviet-American Partial Differen. Equations, Novosibirsk (1963), pp. 3-11 (with M. Sh. Birman).

[177] On some new investigations on the perturbation theory of linear operators, Proc. 4th All-Union Math. Conf., Vol. 1 (1963), 133-134.

[178] On the spectral function of a self-adjoint operator in a

space with indefinite metric, Dokl. Akad. Nauk SSSR **152** (1963), 39-42 (with H. Langer); English transl.: Soviet Math. Dokl. **4** (1963), 1236-1239.

[179] On the theory of quadratic pencils of self-adjoint operators, Dokl. Akad. Nauk SSSR **154** (1964), 1258-1261 (with H. Langer); English transl.: Soviet Math. Dokl. **5** (1964), 266-269.

[180] Criteria for the completeness of the system of root vectors of a contraction, Ukr. Mat. Zh., **16** no.1 (1964), 78-82 (with I.C. Gohberg); English transl.: Amer. Math. Soc. Transl. (2) **54** (1966), 119-124.

[181] Lectures on the Theory of Stability of Solutions of Differential Equations in Banach Space, Naukova Dumka, Kiev (1964).

[182] On certain new studies in the perturbation theory for self-adjoint operators, Proc. First Math. Summer School, Kanev, Vol. 1, Naukova Dumka, Kiev (1964), 103-187; English transl.: Topics in Differential and Integral Equations and Operator Theory, Operator Theory: Advances and Applications, Vol. 7, Birkhäuser, Basel (1983), pp. 107-172.

[183] Some new investigations on the theory of nonselfadjoint operators, Proc. 4th All-Union Math. Congr., Vol. 2 (1964), 261-271 (with M.S. Brodskii, I.C. Gohberg, and V.I. Matsaev); English transl.: Amer. Math. Soc. Transl. (2) **65** (1967), 237-251.

[184] On the factorization of operators in Hilbert space, Acta. Sci. Math. Szeged, **25**, 1-2 (1964), 90-123 (with I.C. Gohberg); English transl.: Amer. Math. Soc. Transl. (2) **51** (1966), 155-188.

[185] On a new application of the fixed point principle in the theory of operators in a space with indefinite metric, Dokl. Akad. Nauk SSSR **154** (1964), 1023-1026; English transl.: Soviet Math. Dokl. **5** (1964), 224-228.

[186] Introduction to the geometry of indefinite J-spaces and to the theory of operators in these spaces, Proc. Second Math. Summer School, Vol. 1, Naukova Dumka, Kiev (1965), 15-92; English transl.: Amer. Math. Soc. Transl. (2) **93** (1970), 103-176.

[187] Introduction to the Theory of Linear Non-Selfadjoint Operators in Hilbert Space, Nauka, Moscow (1965) (with I.C. Gohberg); English transl.: Transl. Math. Monographs, Vol. 18, Amer. Math. Soc., Providence, R.I. (1970).

[188] On the theory of weighted integral equations, Izv. Akad. Nauk Mold. SSR **7** (1965), 40-46.

[189] On a multiplicative representation of the characteristic functions of operators close to unitary ones, Dokl. Akad. Nauk SSSR **161** (1965), 732-735 (with I.C. Gohberg); English transl.: Soviet Math. Dokl. **6** (1965), 1279-1283.

[190] On some mathematical principles of the linear theory of damped oscillations of continua, Proc. Internat. Sympos., Tbilisi, Appl. of Function Theory in Mech. of Continua, Vol. 2 (1965), 283-322 (with H. Langer); English transl. Integral Equations and Operator Theory 1 no.3 (1978), 364-399, and 1 no.4 (1978), 539-566.

[191] Spectral theory of operators in spaces with indefinite metric. II, Letter to the editor, Tr. Mosk. Mat. O.-va **15** (1966), 452-454 (with I.S. Yokhvidov).

[192] On some new Banach algebras and Wiener-Lévy type theorems for Fourier series and integrals, Mat. Issled., 1, no.1 (1966), 82-109; English transl.: Amer. Math. Soc. Transl. (2) 93 (1970), 177-199.

[193] On some new results in the theory of resolvents of Hermitian operators, Dokl. Akad. Nauk SSSR **169** (1966), 1269-1272 (with Sh. N. Saakyan); English transl.: Soviet Math. Dokl. **7** (1966), 1086-1089.

[194] Plus-operators in a space with indefinite metric, Mat. Issled., 1 no. 1 (1966), 131-161 (with Yu. L. Shmul'yan); English transl.: Amer. Math. Soc. Transl. (2), **85** (1969), 93-113.

[195] On the distribution of roots of polynomials orthogonal on the unit circle with respect to a sign-alternating weight, in: Function Theory, Function. Anal. and their Appl., no.2 (1966), 131-137.

[196] On a class of operators in a space with indefinite metric, Dokl. Akad. Nauk SSSR **171** (1966), 34-37 (with Yu. L. Shmul'yan); English transl.: Soviet Math. Dokl. **7** (1966), 1137-1141.

[197] J-polar representation of plus operators, Mat. Issled., 1, no.2 (1966), 172-210 (with Yu. L. Shmul'yan); English transl.: Amer. Math. Soc. Transl. (2) **85** (1969), 115-143.

[198] On triangular representations of linear operators and multiplicative representations of their characteristic functions, Dokl. Akad. Nauk SSSR **175** (1967), 272-275 (with I.C. Gohberg); English transl.: Soviet Math. Dokl. **8** (1967), 831-834.

[199] On a description of contraction operators similar to unitary ones, Fukts. Anal. Prilozhen., **1**, no.1 (1967), 38-60 (with I.C. Gohberg).

[200] The description of all solutions of the truncated power moment problem and some problems of operator theory, Mat. Issled., **2** (1967), 114-132; English transl.: Amer. Math. Soc. Transl. (2) **95** (1970), 219-234.

[201] Theory and Applications of Volterra Operators in Hilbert Space, Nauka, Moscow (1967) (with I.C. Gohberg); English transl.: Transl. Math. Monographs, Vol. 24, Amer. Math. Soc., Providence, R.I. (1970).

[202] On linear-fractional transformations with operator coefficients, Mat. Issled., **2**, no.3 (1967), 64-96 (with Yu. L. Shmul'yan): English transl.: Amer. Math. Soc. Transl. (2) **103** (1974), 125-152.

[203] Analytic problems and results in the theory of linear operators in Hilbert space, Proc. Internat. Congress Mathematicians, Moscow, 1966, Mir, Moscow (1968), 189-216; English transl.: Amer. Math. Soc. Transl. (2) **70** (1968), 68-72 and (2) **90** (1970), 181-209.

[204] R-functions: analytic functions mapping the upper half-plane into itself, Appendix I to the Russian transl. of F.V. Atkinson's book "Discrete and Continuous Boundary Problems", Mir, Moscow (1968), 629-647 (with I.S. Kats); English transl.: Amer. Math. Soc. Transl. (2), **103** (1974), 1-18.

[205] On the spectral functions of the string, Appendix II to the Russian transl. of F.V. Atkinson's book "Discrete and Continuous Boundary Problems", Mir, Moscow (1968), 648-737 (with I.S. Kats); English transl.: Amer. Math. Soc. Transl. (2) **103** (1974), 19-102.

[206] Infinite Hankel matrices and the generalized problems of Carathéodory-Fejér and R. Riesz, Funkts. Anal. Prilozhen., **2**, no. 1 (1968), 1-19 (with V.M. Adamyan and D.Z. Arov); English transl.: Function. Anal. Appl., **2**, no.1 (1968), 1-18.

[207] Infinite Hankel matrices and the generalized problems of Carathéordory-Féjer and Schur, Funkts. Anal. Prilozhen., **2**, no.4 (1968), 1-17 (with V.M. Adamyan and D.Z. Arov); English transl.: Function. Anal. Appl., **2**, no.4 (1968), 269-281.

[208] On the theory of S-matrices of canonical differential

equations with summable potential, Dokl. Akad. Nauk Arm. SSR **46**, no.4 (1968), 150-155 (with F.E. Melik-Adamyan); English transl.: Topics in Differential and Integral Equations and Operator Theory, Operator Theory: Advances and Appl., Vol. 7, Birkhäuser, Basel (1983), 295-302.

[209] Bounded operators commuting with a contraction of class C_{00} whose rank of unitarity is one, Funkts. Anal. Prilozhen., 3, no.3 (1969), 86-87 (with V.M. Adamyan and D.Z. Arov); English transl.: Function. Anal. Appl., **3**, no.3 (1969), 242-243.

[210] General theorems on triangular representations of linear operators and multiplicative representations of their characteristic functions, Funkts. Anal. Prilozhen., 3, no.4 (1969), 1-27 (with V.M. Brodskii and I.C. Gohberg); English transl.: Function, Anal. Appl., **3**, no.4 (1969), 255-276.

[211] Chebyshev-Markov inequalities in the theory of the spectral function of a string, Mat. Issled., 5, no.1 (1970), 77-101; English transl.: Amer. Math. Soc. Transl. (2) **103** (1974), 103-124.

[212] Definition and basic properties of the characteristic function of a G-node, Funkts. Anal. Prilozhen., **4**, no.1 (1970), 88-90 (with V.M. Brodskii and I.C. Gohberg); English transl.: Function. Anal. Appl. **4**, no.1 (1970), 78-80.

[213] The resolvent matrix of a Hermitian operator and the characteristic functions connected with it, Funkts. Anal. Prilozhen., **4**, no.3 (1970), 103-104 (with Sh. N. Saakyan); English transl.: Function. Anal. Appl., **4**, no.3 (1970), 258-259.

[214] Some applications of the theorem on the factorization of a unitary matrix, Funkts. Anal. Prilozhen., **4**, no.4 (1970), 73-75 (with F.E. Melik-Adamyan) English transl.: Function. Anal. Appl. **4**, No. 4 (1970), 327-329.

[215] Some results and problems in the theory of stability of differential equations in a Banach space, Proc. 5th Internat. Conf. Non-linear Oscillations, Kiev, 1969, Vol. I, Akad. Nauk Ukr. SSR (1970), 332-347 (with Yu. L. Daletskii).

[216] Stability of Solutions of Differential Equations in Banach Space, Nauka, Moscow (1970) (with Yu. L. Daletskii); English transl.: Transl. Math. Monographs, Vol. 43, Amer. Math. Soc., Providence, R.I. (1974).

[217] New inequalities for the characteristic numbers of integral equations with smooth kernels, Mat. Issled., **5**, no. 1 (1970), 22-39 (with I.C. Gohberg); English transl.: Topics in Differential and Integral Equations and Operator Theory, Operator Theory: Advances and Appl., Vol. 7, Birkhäuser, Basel (1983), 273-293.

[218] Analytic properties of the Schmidt pairs of a Hankel operator and the generalized Schur-Takagi problem. Mat. Sb., **85**, no.1 (1971), 39-73 (with V.M. Adamyan and D.Z. Arov); English transl.: Math. USSR, Sb., **15** (1971), 31-73.

[219] Infinite Hankel block-matrices and related extension problems, Izv. Akad. Nauk Arm. SSR Ser. Mat., **6**, No. 2-3 (1971), 87-112 (with V.M. Adamyan and D.Z. Arov); English transl.: Amer. Math. Soc. Transl. (2) **111** (1978) 133-156.

[220] Hankel operators, problems of extrapolation of matrix-functions, and the theory of S-matrices, All-Union Conf. Theory of Functions of a Complex Variable, Khar'kov (1971), 6-8 (with V.M. Adamyan and D.Z. Arov).

[221] On the characteristic functions of an invertible operator, Acta Sci. Math. Szeged, **32**, no. 1-2 (1971), 141-164 (with V.M. Brodskii and I.C. Gohberg).

[222] The defect subspaces and generalized resolvents of a Hermitian operator in a space Π_κ, I, Funkts. Anal. Prilozhen., **5**, no. 2 (1971), 59-71 (with H. Langer); English transl.: Function. Anal. Appl., **5**, no. 1 (1971), 136-146.

[223] The defect subspaces and generalized resolvents of a Hermitian operator in a space Π_κ. II, Funkts. Anal. Prilozhen., **5**, no. 3 (1971), 54-69 (with H. Langer); English transl.: Function. Anal. Appl., **5**, no. 3 (1971), 217-228.

[224] Über die verallgemeinerten Resolventen und die charackterische Funktion eines isometrischen Operators im Raume Π_κ, Coll. Math. Soc. János Bolyai, **5**, Hilbert Space Operators and Operator Algebras (1971), 353-400 (with H. Langer).

[225] The Markov Moment Problem and Extremal Problems, Nauka, Moscow (1973) (with A.A. Nudel'man); English transl.: Transl. Math. Monographs, Vol. 50, Amer. Math. Soc., Providence, R.I. (1977).

[226] On some new problems for Hardy-class functions and continuous families of doubly orthogonal functions, Dokl.

Akad. Nauk SSSR, **209** (1973), 537-540 (with P.Ya. Nudel'man); English transl.: Soviet Math. Dokl. **14** (1973), 435-439.

[227] On a conjecture of A.M. Lyapunov, Funkts. Anal. Prilozhen., **7**, no.3 (1973), 45-54; English transl.: Function. Anal. Appl., **7**, no.3 (1973), 204-212.

[228] Uber die Q-Funktionen eines Π -Hermiteschen Operators im Raume Π_κ, Acta Sci. Math. Szeged, **34** (1973), 191-230 (with H. Langer).

[229] On some spatial isoperimetric problems, Kvant, No.2 (1973), 22-26 (with A.A. Nudel'man).

[230] A ballistic problem in cosmos, Kvant, No.5 (1973), 2-6 (with K.R. Kovalenko).

[231] Approximation of functions in $L_2(w_1,w_2)$ by transfer functions of linear systems with minimal energy, Problemy Pered. Inform., **11**, no.2 (1975), 37-60 (with P. Ya. Nudel'man); English transl.: Problems of Inform. Transm. (1975), 124-142.

[232] On the indefinite power moment problem, Dokl. Akad. Nauk SSSR **226** (1976), 261-264 (with H. Langer; English transl.: Soviet Math. Dokl. **17** (1976), 90-93.

[233] On generalized resolvents and resolvent matrices of positive Hermitian operators, Dokl. Akad. Nauk SSSR, **213** (1976), 1063-1066 (with I.E. Ovcharenko); English transl.: Soviet Math. Dokl., **17** (1976), 1705-1709.

[234] On nonlinear integral equations which play a role in the theory of Wiener-Hopf equations. I, Mat. Issled., no.42 (1976), 47-90 and II, Mat. Issled., no. 45 (1977), 67-92; English transl.: Topics in Differential and Integral Equations and Operator Theory, Operator Theory: Advances and Appl., Vol. 7, Birkhäuser, Basel (1983), 173-242.

[235] On the theory of generalized resolvents of non-densely defined Hermitian contractions, Dokl. Akad. Nauk Ukr.SSR, Ser. A, No. 10 (1977), 881-884 (with I.E. Ovcharenko). (Ukrainian)

[236] On Q-functions and sc-resolvents of non-densely defined Hermitian contractions, Sib. Mat. Zh., **18**, no. 5 (1977), 1032-1056 (with I.E. Ovcharenko); English transl.: Sib. Math. J., **18**, no. 5 (1977), 728-746.

[237] Remark of a theorem of V.A. Yakubovic: "Frequency theorem for the case when...", Sib. Mat. Zh., **18**, no.6 (1977), 1411-1412; English transl.: Sib. Math. J., **18**, No.6 (1977), 1001-1002.

[238] On some new results in the theory of factorization of matrix-functions on the unit circle, Dokl. Akad. Nauk SSSR, **234** (1977), 287-290 (with I.M. Spitkovskii); English transl.: Soviet Math. Dokl., **18** (1977), 641-645.

[239] On an interpolation problem connected with the Stieltjes moment problem, Dokl. Akad. Nauk Ukr. SSR, Ser. A., no. 12 (1977), 1068-1072 (with A.A. Nudel'man).

[240] Über einige Fortsetzungsprobleme, die eng mit Hermiteschen Operatoren in Raume Π_κ zusammenhangen. I, Einige Funktionnenklassen und ihre Darstellungen, Math. Nachricht., **77** (1977), 187-236 (with H. Langer).

[241] On measurable Hermitian-positive functions, Mat. Zametki **23**, no. 1 (1978), 79-89; English transl.: Math. Notes **23**, no. 1 (1978), 45-50.

[242] On the factorization of α-sectorial matrix-functions on the unit circle, Mat. Issled., no. 47 (1978), 41-63 (with I.M. Spitkovskii).

[243] Über einige Fortsetzungsproblem, die engt mit Hermiteschen Operatoren in Raume Π_κ zusammenhangen. II, Verallgemeinerte Resolventen, u-Resolventen und ganze Operatoren, J. Functional Anal., **30** (1978), 390-447 (with H. Langer).

[244] On some extension problems which are closely related to the theory of Hermitian operators in a space Π_κ. III. Indefinite analogues of the Hamburger and Stieltjes moment problems, Part I, Beitrage Anal. no. 14 (1979), 25-40 (with H. Langer).

[245] Some propositions on the Fourier-Plancherel and Paley-Wiener type obtained by methods in the theory spectral of functions, Funkts. Anal. Prilozhen , **13**, no.4 (1979), 79-80 (with A.A. Nudel'man); English transl.: Function. Anal. Appl., **13**, no.4 (1979), 301-303.

[246] On direct and inverse problem for the boundary-dissipation frequencies of a non-uniform string, Dokl. Akad. Nauk SSSR, **247** (1979), 1046-1049 (with A.A. Nudel'man); English transl.: Soviet Math. Dokl. **20** (1979), 838-841.

[247] On some extension problems which are closely related to the theory of Hermitian operators in a space Π_κ. III. Indefinite analogues of the Hamburger and Stieltjes moment problems, Part II, Beitrage Anal. No. 15 (1980), 27-45 (with H. Langer).

[248] Stable plus-operators in 3-spaces, Linear Operators, Mat. Issled., no. 54 (1980), 67-83 (with Yu.L.Shmul'yan).

[249] Some propositions on analytic matrix-functions connected with the theory of operators in a space Π_k, Acta. Sci. Szeged, **43** (1981), 181-205 (with H. Langer).

[250] Continual analogues of orthogonal polynomials with respect to an indefinite weight on the unit circle and extension problems related to them, Dokl. Akad. Nauk,, SSSR, **258** (1981), 537-542 (with H. Langer); English transl.: Soviet Math. Dokl., **23** (1981), 553-557.

[251] On representation of entire functions positive on the real axis, or on a semi-axis, or outside a finite interval, Linear Operators and Integral Equations, Mat. Issled., no. 61 (1981), 40-59 (with A.A. Nudel'man); English transl.: Amer. Math. Soc. Transl. (2), 127 (1986), 17-32.

[252] The problem of search of minimum of entropy in indeterminate extension problems, Funkts. Anal. Prilozhen., **15**, no.2 (1981), 61-64 (with D.Z. Arov): English transl.: Function. Anal. Appl., **15**, no. 2 (1981), 123-126.

[253] Spectral shift functions that arise in perturbations of a positive operator, J. Operator Theory, **6**, no.1 (1981), 155-191 (with V.A. Yavryan).

[254] On nonstandard variational problems of determination of optimal form of a vessel, Proc. Anniversary Scientific Session, Bolgarian Institute of Ship Hydrodynamics, Sofia (1981), pp. 77-1 - 77-5 (with V.G. Sizov).

[255] Remarkable limits generated by classical means, Kvant, No. 9, (1981), 13-15 (with A.A. Nudel'man).

[256] Introduction to the Spectral Theory of Operators in Spaces with Indefinite Metric, Math. Research, vol. 9, Akademie-Verlag, Berlin (1982) (with I.S. Iohvidov and H. Langer).

[257] On the theory of inverse problems for canonical differential equations, Dokl. Akad. Nauk Ukr. SSR, Ser. A, No.2 (1982), 14-18 (with I.E. Ovcharenko). (Ukrainian)

[258] Wiener-Hopf equations whose kernels admit an integral representation in terms of exponentials, Izv. Akad. Nauk Arm. SSR, **17**, no.4 (1982), 307-327 and **17**, no.5 (1982), 335-375 (with Yu. L. Shmul'yan); English transl.: Soviet J. Contemp. Math. Anal., **17**, no.4 (1982), 71-93 and **17**, no. 5 (1982), 1-42.

[259] On the calculation of entropy functionals and of their minima in indetermined continuation problems, Acta Sci. Math. Szeged, **45** (1983), 33-50 (with D.Z. Arov).

[260] Some generalizations of Szegö's first limit theorem, Anal. Mat. **9**, no.1 (1983), 23-41 (with I.M. Spitkovskii).

[261] The Borsuk-Ulam theorem, Kvant No. 8 (1983), 20-25 (with A.A. Nudel'man).

[262] Topics in Differential and Integral Equations and Operator Theory, Operator Theory: Adv. Appl., Vol. 7, Birkhäuser Verlag, Basel (1983)

[263] Integral Hankel operators and related continuation problems, Izv. Akad. Nauk Arm. SSR, Mat., **19**, no. 4 (1984), 311-332 and **19**, no.5 (1984), 339-360 (with F.E. Melik-Adamyan): English transl.: Soviet J.Contemp. Math. Anal., **19**, no.4 (1984), 47-68 and **19**, no.5 (1984), 1-22.

[264] On the special representation of a polynomial positive on a system of closed intervals, Preprint 28-84 Fiz. Tekn Inst. Nizkikh Temper. Akad. Nauk Ukr. SSR, Khar'kov (1984) (with B. Ya. Levin and A.A. Nudel'man).

[265] Some function theoretic problems connected with the theory of spectral measures of isometric operators, Lecture Notes,. Math., Vol. 1043 (1984), pp. 160-163 (with V.M. Adamyan and D.Z. Arov).

[266] Approximation of bounded functions by elements of $H^\infty + C$, Lecture Notes Math., Vol. 1043 (1984), pp. 254-256 (with V.M. Adamyan and D.Z. Arov).

[267] Some problems connected with the Szego limit theorems, Lecture Notes Math., Vol. 1043 (1984), (with I.M. Spitkovskii).

[268] Banach algebras of functions generated by the set of all almost periodic polynomials whose exponents belong to a given interval, Lecture Notes Math., Vol. 1043 (1984), pp. 632-635.

[269] On some continuation problems which are closely related to the theory of operators in spaces Π_κ. IV. Continuous analogues of orthogonal polynomials on the unit circle with respect to an indefinite weight and related continuation problems for some classes of functions, J. Operator Theory, **13** (1985), 299-417 (with H. Langer).

[270] A diophantine equation of academician A.A. Markov, Kvant No. 4, (1985), 13-16, and No. 5 (1985), 59.

[271] Matrix-continual analogues of the Schur and Caratheodory-Toeplitz problems, Izv. Akad. Nauk Arm. SSR, 21, no.2 (1986), 107-141 (with F.E. Melik-Adamyan); English Transl.: Soviet J. Contemp. Math. Anal., 21, no.2 (1986).

[272] A supplement to the paper of M.G. Krein and Yu. L. Shmul'yan, "Wiener-Hopf equations whose kernels admit an integral representation in terms of exponentials," Izv. Akad. Nauk. Arm. SSR, Mat., 21 (1986) No.3 (with Yu. L. Shmul'yan).

[273] On perturbation determinants and a trace formula for certain classes of operator pairs. J. Operator Theory, 17, no. 1 (1987), 129-188.

Operator Theory:
Advances and Applications, Vol. 34

ON ORTHOGONAL MATRIX POLYNOMIALS

Daniel Alpay and Israel Gohberg

This paper contains a generalization of M.G. Krein's theorem about the distribution of zeros of orthogonal polynomials for the matrix-valued case. The proof is based on the theory of rational matrix-valued functions unitary on the unit circle.

I. INTRODUCTION.

The aim of this paper is to prove a theorem on the distribution of zeros of orthogonal matrix polynomials. This theorem is a generalization of a theorem of M.G. Krein ([K]; see also [EGL]) which describes the localization of the zeros of scalar orthogonal polynomials with respect to an indefinite weight function. The proof is based on some results about realization and factorization of rational matrix-valued functions unitary on the unit circle which are obtained in [AG2]. To make the paper self-contained, we present the latter results in the second section. The third section contains the formulation of M.G. Krein's theorem. The fourth section contains first properties of matrix-valued orthogonal polynomials, and in the last section the main result is obtained. The results of this paper appeared earlier in the preprint [AG1].

We now review some properties of matrix-valued rational functions; we refer to the monograph [BGK] for more information on these functions. Let W be a rational matrix-valued function analytic at infinity. Then W admits a realization of the form

$$W(\lambda) = D + C(\lambda I_n - A)^{-1}B \tag{1.1}$$

where A, B, C and D are matrices of suitable sizes and I_n denotes the identity in the space $C_{n\times n}$ of $n \times n$ matrices with complex entries. The smallest n in the realization of $W(\lambda)$ in the form (1.1) is called the MacMillan degree of W. If in (1.1) n is equal to the MacMillan degree of W, then the realization (1.1) is called minimal. The realization is minimal if and only if the following two conditions hold: the pair (C, A) is observable, that is,

$$\bigcap_{k=0}^{\infty} \ker\ CA^k = \{0\}$$

and the pair (A, B) is controllable, that is,

$$\text{Span} \bigcup_{k=0}^{\infty} \text{Im}\ BA^k = C_n \ ,$$

where C_n denotes the space of n dimensional column vectors with complex components.

The matrices A, B and C in (1.1) are not uniquely defined. However, when the realization is minimal, they are unique up to similarity: if (1.1) is a minimal realization of W and $W(\lambda) = D + C_1(\lambda I_n - A_1)^{-1}B_1$ is another minimal realization, then there exists a unique invertible matrix S such that

$$C = C_1 S \ , \qquad A = S^{-1}A_1 S \ , \qquad B = S^{-1}B_1 \ .$$

The matrix S is given by each of the formulas

$$S = \left[\text{col}\left(C_1 A_1^j\right)_0^{n-1}\right]^+ \left[\text{col}\left(CA^j\right)_0^{n-1}\right]$$

and

$$S = \left[\text{row}\left(A_1^j B_1\right)_0^{n-1}\right] \left[\text{row}\left(A^j B\right)_0^{n-1}\right]^\dagger$$

where the symbol $+$ indicates a left inverse and the symbol $\dagger$ indicates a right inverse.

Let $W(\lambda) = D + C\left(\lambda I_n - A\right)^{-1} B$ be a realization of the rational matrix function W and suppose D is invertible. Then

$$W^{-1}(\lambda) = D^{-1} - D^{-1}C\left(\lambda I_n - A^\times\right)^{-1} BD^{-1}$$

is a realization of the function W^{-1}, where

$$A^\times = A - BD^{-1}C$$

If one of these realizations is minimal, then the other is also minimal. The operator $A^\times$ plays an important role in the description of all minimal factorizations of W, i.e., of all representations $W = W_1 W_2$, where W_1 and W_2 are rational $C_{m\times m}$ valued functions invertible at infinity such that

$$\deg W = \deg W_1 + \deg W_2$$

where $\deg W$ denotes the MacMillan degree of W (see [BGK], p.83). We now recall this description. A projection π is called a supporting projection if

$$\text{Ker } \pi = M \quad \text{and} \quad \text{Im } \pi = N$$

where M and N are such that

$$\begin{array}{cc} M + N = C_n, & M \cap N = \{0\} \\ AM \subset M, & A^\times N \subset N \end{array}$$

Let $D = D_1 D_2$ be a decomposition of D into two invertible matrices and let π be a supporting projection. Then $W = W_1 W_2$ is a minimal factorization of W, with

$$W_1(\lambda) = D_1 + C(\lambda I_n - A)^{-1}(I_n - \pi) B D_2^{-1}$$

$$W_2(\lambda) = D_2 + D_1^{-1} C \pi (\lambda I_n - A)^{-1} B$$

and any minimal factorization can be obtained in such a way. Moreover, for a given decomposition $D = D_1 D_2$ there is a one to one correspondence between supporting projections and minimal factorizations of W.

Finally, $\mathbb{D}$ will denote the open unit disk, A^* will denote the adjoint of the matrix A, and $F(\lambda)^*$ will denote $(F(\lambda))^*$.

2. RATIONAL MATRIX-VALUED FUNCTIONS UNITARY ON THE UNIT CIRCLE.

In this section, we review the main aspects of the realization and factorization theory of rational matrix-valued functions unitary on the unit circle. The results are adapted from [AG2] and some of the proofs are only outlined. A $C_{m \times m}$ valued rational matrix function U will be called unitary on the unit circle if, at any point z with $|z| = 1$ at which U is defined, $U(z)U^*(z) = I_m$. By analytic continuation this identity extends to any regular point of U as

$$U(z)U^*(1/z^*) = I_m \tag{2.1}$$

Using the Smith form ([GLR] p.193), one can easily deduce the following lemma from equation (2.1).

LEMMA 2.1. *Let U be a rational matrix-valued function unitary on the unit circle. Then U is analytic and invertible at the the point z (including $z = \infty$) if and only if U is analytic and invertible at the point $1/z^*$. Moreover, if z is a singular point of U with partial multiplicities*

$$k_1 \geq k_2 \geq \cdots k_r > 0 \geq k_{r+1} \geq k_{r+2} \geq \cdots \geq k_s$$

then $1/z^*$ *is a singular point of* U *with partial multiplicities*

$$-k_s \geq -k_{s-1} \geq \cdots \geq 0 > -k_r \geq \cdots \geq -k_2 \geq -k_1$$

In the next theorem we characterize the minimal realizations of matrix-valued rational functions unitary on the unit circle and analytic and invertible at infinity.

THEOREM 2.1. *Let* U *be a* $C_{m\times m}$ *valued function analytic and invertible at infinity and let* $U(z) = D + C(zI_n - A)^{-1}B$ *be a minimal realization of* U. *Then* U *is unitary on the unit circle if and only if the following conditions hold*

a) U *is analytic and invertible at the origin.*

b) *There exists an hermitian matrix* H *such that*

$$\begin{pmatrix} A & B \\ C & D \end{pmatrix}^* \begin{pmatrix} H & 0 \\ 0 & I_m \end{pmatrix} \begin{pmatrix} A & B \\ C & D \end{pmatrix} = \begin{pmatrix} H & 0 \\ 0 & I_m \end{pmatrix} \tag{2.2}$$

The hermitian matrix H *in equality (2.2) is invertible.*

PROOF. The necessity of condition a) follows from Lemma 2.1. We now prove the necessity of b). Equation (2.1) leads to

$$U^{-1}(z) = (U(\frac{1}{z^*}))^*$$

and hence, with $A^\times = A - BD^{-1}C$,

$$D^{-1} - D^{-1}C(zI_n - A^\times)^{-1}BD^{-1} = D^* + zB^*(I_n - zA^*)^{-1}C^*$$

Setting $D_1 = D^* - B^*(A^*)^{-1}C^*$, we can rewrite this equation as

$$D^{-1} - D^{-1}C(zI_n - A^\times)^{-1}BD^{-1} = D_1 - B^*(A^*)^{-1}(zI_n - (A^*)^{-1})^{-1}(A^*)^{-1}C^* \tag{2.3}$$

Letting z go to infinity, we get $D_1 = D^{-1}$, i.e.,

$$D^{-1} = D^* - B^*(A^*)^{-1}C^* \tag{2.4}$$

Moreover, equation (2.3) is an equality between two minimal realizations of a given rational function and thus there exists a unique invertible matrix H such that

$$D^{-1}C = B^*(A^*)^{-1}H \tag{2.5}$$

$$(A^*)^{-1} = A^\times \tag{2.6}$$

$$H^{-1}(A^*)^{-1}C^* = BD^{-1} \tag{2.7}$$

Equations (2.5) and (2.7) are also satisfied by the matrix H^* and hence $H = H^*$. Moreover, equations (2.4)-(2.7) are easily shown to be equivalent to equation (2.2).

Conversely, let $U(z) = D + C(zI_n - A)^{-1}B$, where A, B, C and D satisfy equation (2.2) for some hermitian matrix H. A short computation leads to

$$U^*(\omega)U(z) = I_m - (1 - z\omega^*)B^*(\omega^* I_n - A)^{-1}H(zI_n - A)^{-1}B \tag{2.8}$$

and hence U is unitary on the unit circle. The matrix H satisfies in particular the Stein equation

$$H - A^*HA = -C^*C \tag{2.9}$$

and is invertible. Details may be found in [AG2], Section 5. □

From the proof of the theorem it follows that the matrix H is uniquely defined; it is called the *associated hermitian matrix* of the minimal realization of the matrix-valued function $U(z) = D + C(zI_n - A)^{-1}B$, which is unitary on the unit circle. The matrix H can be computed as

$$H = (\mathrm{col}(B^*((A^*)^{-1})^{(j-1)})_0^{n-1})^+(\mathrm{col}(D^{-1}C(A^\times)^j)_0^{n-1})$$

or

$$H = (\mathrm{row}(((A^*)^{-1})^{j-1}C^*)_0^{n+1})(\mathrm{row}((A^\times)^j BD^{-1})_0^{n+1})^\dagger$$

where $+$ (resp $\dagger$) denotes a left (resp a right) inverse.

We note that condition b) of the theorem is equivalent to

b') There exists an hermitian matrix G such that

$$\begin{pmatrix} A & B \\ C & D \end{pmatrix} \begin{pmatrix} G & 0 \\ 0 & I_m \end{pmatrix} \begin{pmatrix} A & B \\ C & D \end{pmatrix}^* = \begin{pmatrix} G & 0 \\ 0 & I_m \end{pmatrix} \tag{2.10}$$

This matrix G is invertible.

We also note the formula

$$U(z)U^*(\omega) = I_m - (1 - z\omega^*)C(zI_n - A)^{-1}H^{-1}(\omega^* I_n - A^*)^{-1}C^* \tag{2.11}$$

which is valid for any regular points z and ω of A.

The next theorem concerns the problem of reconstructing a matrix-valued rational function unitary on the unit circle from its left pole structure.

THEOREM 2.2. *Let (C, A) be an observable pair of matrices with A invertible. Then there exists a rational function unitary on the unit circle with minimal realization $U(z) = D + C(zI_n - A)^{-1}B$ if and only if the Stein equation*

$$H - A^*HA = -C^*C \tag{2.9}$$

has an hermitian solution. If such a solution exists, it is invertible, and possible choices of B and D are given by

$$D_\alpha = I_m + CH^{-1}(I_n - \alpha A^*)^{-1}C^* \tag{2.12}$$

$$B_\alpha = H^{-1}(A^*)^{-1}C^*D_\alpha \tag{2.13}$$

where α with $|\alpha| = 1$ is a regular point of A. For a given H, any other choices of B and D differ from B_α and D_α by a right unitary multiplicative constant matrix.

PROOF. Let H be an hermitian solution of the Stein equation (2.9). Then as was mentioned in the proof of Theorem 2.1, H is invertible. Moreover, for α with $|\alpha| = 1$ in the resolvent set of A, the pair (A, B_α) is easily seen to be controllable; thus, $U_\alpha(z) = D_\alpha + (zI_n - A)^{-1}B_\alpha$ is a minimal realization of a matrix-valued rational function U_α. The matrix $\begin{pmatrix} A & B_\alpha \\ C & D_\alpha \end{pmatrix}$ satisfies equation (2.2), and Theorem 2.2 permits us to conclude that U_α is unitary on the unit circle.

Let B and D be another solution of the inverse problem, for a given H, and let $U(z) = D + C(zI_n - A)^{-1}B$. Formula (2.11) leads to

$$U(z)U^*(\omega) = U_\alpha(z)U_\alpha^*(\omega)$$

Hence the functions U and U_α differ by a unitary multiplicative constant, and the final conclusion follows therefore from the observability of the pair (C, A). □

Different minimal realizations of a given rational matrix-valued function unitary on the unit circle have in general different associated hermitian matrices. The following lemma describes the connections between them.

LEMMA 2.2. *Let U be a rational function, analytic and invertible at infinity and unitary on the unit circle. Let $U(z) = D + C_i(zI_n - A_i)^{-1}B_i, i = 1, 2$, be two minimal realizations of U, with associated hermitian matrices $H_i, i = 1, 2$. Then the two minimal realizations are similar, that is,*

$$C_1 = C_2S\ , \qquad A_1 = S^{-1}A_2S\ , \qquad B_1 = S^{-1}B_2 \tag{2.14}$$

for a unique invertible matrix S, and

$$H_1 = S^*H_2S \tag{2.15}$$

In particular, the matrices H_1 and H_2 have the same number of positive and negative eigenvalues.

PROOF. The existence and uniqueness of the similarity matrix S are well known (see e.g. [BGK] p. 65-66). It is easy to check that both H_1 and S^*H_2S satisfy equation (2.2) with A_1, B_1, C_1, D_1 instead of A, B, C, D, and hence (2.15) holds. □

Let $K(z,\omega)$ be a $C_{m\times m}$ valued function, defined for z and ω in some set Ω and such that $K^*(z,\omega) = K(\omega, z)$. It has k negative squares if for any positive integer r, points $\omega_1, \ldots, \omega_r$ in Ω, and vectors $c_1, \ldots c_r$ in C_m the $C_{r\times r}$ hermitian matrix

$$\left(c_j^* K(\omega_j, \omega_i) c_i\right)_{i,j=1}^r$$

has at most k negative eigenvalues and exactly k negative eigenvalues for some choice of $\omega_1, \ldots \omega_r, c_1, \ldots, c_r$ (see [KL]). With this definition, we can now state the following theorem, which gives a characterization of the number of negative eigenvalues of the associated hermitian matrix H.

THEOREM 2.3. *Let U be a rational matrix-valued function unitary on the unit circle and analytic and invertible at infinity. Let $U(z) = D + C(zI_n - A)^{-1}B$ be a minimal realization of U with associated hermitian matrix H. Then the number of negative eigenvalues of H is equal to the number of negative squares of each of the functions*

$$K_U(z,\omega) = \frac{I_m - U(z)U^*(\omega)}{1 - z\omega^*} \quad \text{and} \quad \widehat{K}_U(z,\omega) = \frac{I_m - U^*(\omega)U(z)}{1 - z\omega^*} \tag{2.16}$$

Finally, let $\mathcal{K}(U)$ be the span of the functions $z \to K_U(z,\omega)c$, where ω spans the points of analyticity of U and c spans C_m. Then $\mathcal{K}(U)$ is a finite-dimensional vector space of rational functions and its dimension is equal to the MacMillan degree of U.

PROOF. From formula (2.11) it follows that

$$K_U(z,\omega) = C(zI_n - A)^{-1}H^{-1}(\omega^* I_n - A^*)^{-1}C^* \tag{2.17}$$

Let $\omega_1, \ldots, \omega_r$ be in the resolvent set of A and let $c_1, \ldots, c_r$ be in C_m. Then from the matrix equalities

$$(c_j^* K_U(\omega_j, \omega_i)c_i)_{i,j=1,r} = X^* H^{-1} X$$

$$X = ((\omega_1^* - A^*)^{-1}C^*c_1, \ldots, (\omega_r^* - A^*)^{-1}C^*c_r)$$

it follows that the function K_U has at most k_H negative squares, where k_H denotes the number of negative eigenvalues of the hermitian matrix H. The pair (C, A) is observable, and thus we can choose a basis of C_n of the form $x_i = (\omega_i^* - A_i^*)^{-1}C^*c_i$, $i = 1, \ldots, n$. In particular, det $X \neq 0$ for $X = (x_1, \ldots, x_n)$ and the matrix $X^*H^{-1}X$ has exactly k_H negative squares, and thus K_U has k_H negative squares.

The case of the function $K_{\widehat{U}}(z,\omega)$ is treated similarly, and relies on formula (2.10).

From (2.17) we see that any linear combination of functions $K_U(z,\omega)c$ is of the form

$$C(zI_n - A)^{-1}f$$

for some vector f in C_n, and thus dim $\mathcal{K}(U) \leq m$. The observability of the pair (C, A) leads to the conclusion that

$$C(zI_n - A)^{-1}f \equiv 0 \Rightarrow f = 0$$

and thus to the result that dim $\mathcal{K}(U) = m$. □

We will denote by $\nu(U)$ the number of negative squares of either of the functions defined in (2.16).

THEOREM 2.4. *Let $U_i, i = 1,2$, be two $C_{m\times m}$ valued rational functions unitary on the unit circle and analytic and invertible at infinity, with minimal realizations $D_i + C_i({}_zI_{n_i} - A_i)^{-1}B_i$, $i = 1,2$, and associated hermitian matrices $H_i, i = 1,2$, and suppose the product $U = U_1U_2$ is minimal. Then $U(z) = D + C({}_zI_n - A)^{-1}B$, where $n = n_1 + n_2$,*

$$D = D_1D_2\ , \quad C = (C_1 \quad D_1C)\ , \quad B = \begin{pmatrix} B_1D_2 \\ B_2 \end{pmatrix} \tag{2.18}$$

and

$$A = \begin{pmatrix} A_1 & B_1C_2 \\ 0 & A_2 \end{pmatrix} \tag{2.19}$$

is a minimal realization of U with associated Hermitian matrix

$$H = \begin{pmatrix} H_1 & 0 \\ 0 & H_2 \end{pmatrix} \tag{2.20}$$

PROOF. It suffices to check that equation (2.2) is satisfied with A, B, C, D and H defined in (2.17)-(2.19). The other assertion is a consequence of Lemma 2.2. □

COROLLARY 2.1. *Let U_1 and U_2 be rational matrix-valued functions unitary on the unit circle and analytic and invertible at infinity, and suppose the factorization $U = U_1U_2$ is minimal. Then*

$$\nu(U_1U_2) = \nu(U_1) + \nu(U_2) \tag{2.21}$$

This additional property of the number of negative squares can be found in [S].

We now study the multiplicative structure of a matrix-valued rational function unitary on the unit circle. First we need some definitions. For an invertible hermitian matrix H we denote by $[,]_H$ the hermitian form

$$[x,y]_H = \langle x, Hy\rangle$$

where $\langle , \rangle$ denotes the usual euclidean inner product of C_n. Two vectors x and y are called H-orthogonal if $[x,y]_H = 0$. For a subspace M of C_n define

$$M^{[\perp]} = \{x \in C_n \ , \ [x,y]_H = 0 \qquad \forall y \in M\}$$

The subspace M is called nondegenerate if no vector of M is H-orthogonal to all of M, that is, if $M \cap M^{[\perp]} = \{0\}$. In such a case,

$$M[\dot{+}]M^{[\perp]} = C_n \tag{2.22}$$

where the symbol $[\dot{+}]$ denotes a direct H-orthogonal sum. The adjoint matrix $A^{[*]}$ of a matrix in $C_{n\times n}$ with respect to $[,]_H$ is defined by

$$[Ax,y]_H = [x, A^{[*]}y]_H \qquad x,y \in C_n$$

and it is easily verified that

$$A^{[*]} = HA^*H^{-1} \tag{2.23}$$

THEOREM 2.5. *Let U be a rational function unitary on the unit circle and analytic and invertible at infinity. Let $U(z) = D + C(zI_n - A)^{-1}B$ be a minimal realization of U with associated hermitian matrix H and associated inner product $[,]_H$. Then*

a) A has no spectrum on the unit circle

b) $[f,f]_H \neq 0$ for any eigenvector f of A

c) U is a minimal product of a unitary constant and of N rational matrix functions of the from

$$I_m - P + \frac{z-\omega}{1-z\omega^*}P \tag{2.24}$$

where P is an orthogonal projection of rank 1 and $|\omega| \neq 1$.

PROOF. The matrix H satisfies the Stein equation

$$H - A^*HA = -C^*C$$

Hence, for f an eigenvector of A corresponding to the eigenvalue μ,

$$\langle Hf,f\rangle - \langle HAf, AF\rangle = \langle Cf, Cf\rangle$$

that is,

$$[f,f]_H(1-|\mu|^2) = -\langle Cf, Cf\rangle$$

Suppose assertion a) or b) is false. Then $Cf = 0$ and hence $CA^n f = 0$ since $A^n f = \mu^n f$. We deduce that f belongs to the intersection $\bigcap_0^\infty \text{Ker } CA^m = \{0\}$. The latter is impossible.

Since $[f,f]_H \neq 0$, the subspace $M = \text{span }\{f\}$ is nondegenerate and (2.22) holds. Moreover, because of (2.23) equation (2.6) can be rewritten as

$$A^\times = (A^{[*]})^{-1}$$

Therefore the pair $(M, M^{[\perp]})$ defines a supporting projection π by

$$\text{Ker } \pi = M \qquad \text{and} \qquad \text{Im } \pi = M^{[\perp]}$$

By the factorization theorem recalled in the introduction ([BGK] p.14), for any factorization $D = D_1D_2$ of D into two invertible matrices, the factorization $U = U_1U_2$ where

$$U_1(z) = (I_m + C(zI_n - A)^{-1}(I_n - \pi)BD^{-1})D_1$$
$$U_2(z) = D_2(I_m + D^{-1}C\pi(zI_n - A)^{-1}B)$$

is a minimal factorization of U into two rational matrix-valued functions.

Let P denote the orthogonal projection from C_n onto M in the usual metric of C_n. The operator $H_1 = PH\big|_M$ is easily seen to be nonzero since $[f,f]_H \neq 0$. Moreover, since

$$(I_n - \pi^*)H = H(I_n - \pi)$$

and

$$P(I_n - \pi^*) = P$$

one sees that $H_1^{-1} = (I_n - \pi)H^{-1}\big|_M$, and thus, taking into account (2.7), one can rewrite the matrix function U_1 as

$$U_1(z) = (I_m + C_1(zI_n - A_1)^{-1}H_1^{-1}(A_1^*)^{-1}C_1^*)D_1$$

where

$$C_1 = C\big|_M \qquad A_1 = A\big|_M$$

and

$$B_1 = (I_n - \pi)B$$

We note that $A_1 = \mu, C_1$ is a vector and by equation (2.9) these quantities are linked by

$$H_1(1 - |\mu|^2) = -C_1^* C_1 \tag{2.25}$$

We choose D_1 of the form

$$D_1 = I_m + C_1 H_1^{-1}(I - \alpha A_1^*)^{-1} C_1^*$$

where α is a regular point of A with $|\alpha| = 1$. The matrix-valued function U_1 can be rewritten as

$$\begin{aligned} U_1(z) &= \left(I_m + \frac{C_1 C_1^*}{H_1(z-\mu)\mu^*}\right)\left(I_m + \frac{C_1 C_1^*}{H_1(1-\alpha\mu^*)}\right) \\ &= I_m - \frac{C_1 C_1^*}{C_1^* C_1} + \frac{C_1 C_1^*}{C_1^* C_1} \cdot \left(\frac{\alpha - \mu}{1 - \alpha\mu^*}\right) \cdot \left(\frac{1 - z\mu^*}{z - \mu}\right) \end{aligned}$$

Setting $\mu = 1/\omega^*, P = C_1 C_1^* / C_1^* C_1$ and choosing of $\alpha = (1+\omega)/1-\omega^*)$, we obtain U_1 in the form (2.24). We can reiterate this procedure since U_2 is itself unitary on the unit circle; after m steps we get a unitary constant matrix. □

A suitable Moebius transformation permits us to get the multiplicative structure when the function U is not analytic or invertible at infinity, and we obtain the following theorem.

THEOREM 2.6. *Let U be a rational matrix-valued function unitary on the unit circle and of MacMillan degree n. Then U is a minimal product of a unitary constant and of n factors of the form (2.13).*

The last theorem of this section is the finite-dimensional version of a more general theorem due to Krein and Langer. This theorem will be used in Section 5 to prove a generalization of M.G. Krein's theorem on the zeros of orthogonal polynomials.

We first need the following definition: A Blaschke-Potapov product is the product of a unitary constant matrix and a finite product of terms of the form (2.24) with ω in $I\!D$. By Theorem 2.5, a rational function unitary on the unit circle is a Blaschke-Potapov product if and only if it is analytic in $I\!D$. The MacMillan degree of a Blaschke-Potapov product is equal to the total multiplicity of its zero.

THEOREM 2.7. *Let U be a rational matrix-valued function unitary on the unit circle. Then there exist Blaschke-Potapov products B_1, B_2, B_3, B_4 such that*

$$U = B_1 B_2^{-1} \tag{2.26}$$

$$U = B_2^{-1} B_4 \tag{2.27}$$

with

$$\deg U = \deg B_1 + \deg B_2 = \deg B_3 + \deg B_4 \tag{2.28}$$

and

$$\nu(U) = \deg B_3 = \deg B_2 \tag{2.29}$$

In particular, $\nu(U)$ *is equal to the number of poles of* U *in* $\mathbb{D}$ *and* $n - \nu(U)$ *is equal to the number of poles of* U *in* $\{z, |z| > 1\}$.

PROOF. We first suppose that U is analytic and invertible at infinity. Then the representations (2.26)-(2.27) follow from the proof of Theorem 2.6, where to get (2.26), at each iteration, one considers first poles outside $\mathbb{D}$, and to get (2.27) one first considers poles inside $\mathbb{D}$. From equation (2.25) we see that the associated hermitian matrix associated with a factor U_1 of the from (2.24) is a positive number when μ is outside $\mathbb{D}$ and a negative number when μ is inside $\mathbb{D}$.. Then the equality $\deg B_3 = \deg B_2 = \nu(U)$ follows from Theorem 2.3 and Theorem 2.4. It is then clear that $\nu(U)$ is equal to the number of poles of U in $\mathbb{D}$, and that $n - \nu(U)$ is equal to the number of poles of U outside $\mathbb{D}$.

The case where U is not analytic or invertible at infinity is treated by first considering $V(z) = U(\phi(z))$, where ϕ is a Moebius transformation which maps $\mathbb{D}$ onto $\mathbb{D}$ and such that V is analytic and invertible at infinity. □

As a corollary of Theorem 2.7 we have

COROLLARY 2.1. *Let* $U = B_1 B_2^{-1}$ *be as in Theorem 2.7. Then there is no non-zero vector* c *such that*

$$B_1(\omega)c = B_2(\omega)c = 0$$

PROOF. Let c be such that $B_1(\omega)c = B_2(\omega)c = 0$. The point ω is in $\mathbb{D}$ since B_1 and B_2 are Blaschke Potapov-products. Thus B_1 and B_2 are analytic at ω and

$$B_i'(z) = B_i(z)\left(I_m - \frac{cc^*}{c^*c} + \frac{1 - z\omega^*}{z - \omega}\frac{cc^*}{c^*c}\right)$$

are still Blaschke-Potapov products for $i = 1, 2$. But $\deg B_i' = \deg B_i - 1$ for $i = 1, 2$ and

$$U = B_1'(B_2')^{-1}$$

with $\deg B_1' + \deg B_2' < \deg U_1$. This is a contradiction. □

3. THE THEOREM OF M.G. KREIN

Let $R_0, R_1, \ldots,$ be a sequence of $C_{p\times p}$ matrices, and let T_N denote the block Toeplitz matrix

$$T_N = \begin{pmatrix} R_0 & R_{-1} & \cdots & R_{-N} \\ R_1 & R_0 & \cdots & R_{1-N} \\ \cdot & \cdot & & \cdot \\ \cdot & \cdot & & \cdot \\ \cdot & \cdot & & \cdot \\ R_N & R_{N-1} & \cdots & R_0 \end{pmatrix} \tag{3.1}$$

where $R_{-i} = R_i^*$. We suppose that the matrix T_N is invertible. Then the orthogonal polynomial associated to T_N is defined by

$$P_N(z) = \Gamma_{0N} + z\Gamma_{1N} + \cdots + z^N\Gamma_{NN} \tag{3.2}$$

where $(\Gamma_{ij})_{i,j=0,\cdots,N}$ is the block decomposition of T_N^{-1} into $C_{p\times p}$ matrices. (These should be denoted by $\Gamma_{ij}^{(N)}$, but the dependence on N is omitted, to lighten the notation).

The distribution of the zeros of P_N with respect to the unit circle is given in a theorem of M.G. Krein ([K]):

THEOREM 3.1. *Suppose the Toeplitz matrices T_k are invertible for $k = 0, \ldots, N$, and let $D_k = \det T_k$. Let $\beta(N)$ (resp. $\gamma(N)$) be the number of permanence (resp. changes) of sign of the sequence*

$$1, D_0, D_1, \ldots, D_{N-1} .$$

If $D_N D_{N-1} > 0$, then P_N has $\beta(N)$ zeros (counting multiplicities) inside the unit circle.

If $D_N D_{N-1} < 0$, then P_N has $\gamma(N)$ zeros inside the unit circle.

The situation is more involved in the non-scalar case. Indeed, take $R_0 = \begin{pmatrix} I_k & 0 \\ 0 & -I_k \end{pmatrix}$ and $R_1 = \begin{pmatrix} 0 & I_k \\ I_k & 0 \end{pmatrix}$, and let

$$T_1 = \begin{pmatrix} R_0 & R_1 \\ R_1 & R_0 \end{pmatrix}$$

Then easy computations show that the associated orthogonal polynomial P_1 is equal to

$$P_1(z) = \frac{1}{2}\begin{pmatrix} I_k & zI_k \\ zI_k & I_k \end{pmatrix}$$

In particular, the zeros of P_1 are $\pm i$, i.e., P_1 has all its zeros on the unit circle. Thus, Theorem 3.1 has no straight-forward generalization to the matrix case, and additional assumptions will be needed to prove a generalization of Theorem 3.1.

4. FIRST PROPERTIES OF ORTHOGONAL POLYNOMIALS

One of the assumptions we make to prove a generalization of Theorem 3.1 is that T_{N-1} is invertible. In this case, the following theorem holds.

THEOREM 4.1. *Let the block Toeplitz matrices T_N and T_{N-1} be invertible. Then the blocks Γ_{00} and Γ_{NN} of T_N^{-1} are invertible and*

$$G(z,\omega) \stackrel{\text{def}}{=} \sum_{i,j=0}^{N} z^i(\omega^*)^j\Gamma_{ij} = \frac{Q_N(z)\Gamma_{00}^{-1}Q_N(\omega)^* - z\omega^* P_N(z)\Gamma_{NN}^{-1}P_N(\omega)^*}{1-z\omega^*} \tag{4.1}$$

where

$$Q_N(z) = \Gamma_{00} + z\Gamma_{10} + \cdots + z^N\Gamma_{N0} \tag{4.2}$$

PROOF. We first recall the following matrix formulas, where A, B, C, D are in $C_{p\times p}$,$C_{p\times q}$,$C_{q\times p}$, and $C_{q\times q}$ respectively.

$$\begin{pmatrix} A & B \\ C & D \end{pmatrix} = \begin{pmatrix} I_p & 0 \\ CA^{-1} & I_q \end{pmatrix}\begin{pmatrix} A & 0 \\ 0 & A^{\square} \end{pmatrix}\begin{pmatrix} I_p & A^{-1}B \\ 0 & I_q \end{pmatrix} \tag{4.3}$$

when A is invertible with

$$A^{\square} = D - CA^{-1}B \tag{4.4}$$

and

$$\begin{pmatrix} A & B \\ C & D \end{pmatrix} = \begin{pmatrix} I_p & BD^{-1} \\ 0 & I_q \end{pmatrix}\begin{pmatrix} A^{\times} & 0 \\ 0 & D \end{pmatrix}\begin{pmatrix} I_p & 0 \\ D^{-1}C & I_q \end{pmatrix} \tag{4.5}$$

if D is invertible, with $A^{\times} = A - BD^{-1}C$.

Applying first (4.3) with $\begin{pmatrix} A & B \\ C & D \end{pmatrix} = T_N$ and $A = T_{N-1}$, we see that $A^{\square}$ is invertible and so is Γ_{NN}, since $\Gamma_{NN} = (A^{\square})^{-1}$, as is easily deduced from (4.3). Similarly, applying (4.5) with $\begin{pmatrix} A & B \\ C & D \end{pmatrix} = T_N$ and $D = T_{N-1}$, we get that $A^{\times}$ is invertible and so is Γ_{00}, since (4.5) implies that $\Gamma_{00} = (A^{\times})^{-1}$. Now we use the Gohberg-Heinig formula which looks as follows

$$(T_N)^{-1} = L(\Gamma_{00},\ldots,\Gamma_{N0})\Gamma_{00}^{-1}L(\Gamma_{00},\ldots,\Gamma_{N0})^*$$

$$- L(0,\Gamma_{0N},\ldots,\Gamma_{N-1,N})\Gamma_{NN}^{-1}L(0,\Gamma_{0N},\ldots,\Gamma_{N-1,N})^*$$

where

$$L(x_0,\ldots,x_N)=\begin{pmatrix} x_0 & 0 & \cdots & 0\\ x_1 & x_0 & \cdots & 0\\ . & . & \cdots & .\\ x_N & x_{N-1} & \cdots & x_0\end{pmatrix}$$

Hence, with $Z=\begin{pmatrix} 0 & 0 & \cdots & 0 & 0\\ I_p & 0 & \cdots & 0 & 0\\ \cdots & . & \cdots & . & .\\ 0 & 0 & \cdots & I_p & 0\end{pmatrix}$, we get

$$\Gamma - Z\Gamma Z^* = \begin{pmatrix}\Gamma_{00}\\ \vdots\\ \Gamma_{N0}\end{pmatrix}\Gamma_{00}^{-1}\begin{pmatrix}\Gamma_{00}\\ \vdots\\ \Gamma_{N0}\end{pmatrix}^* - \begin{pmatrix}0\\ \Gamma_{0N}\\ \vdots\\ \Gamma_{N-1,N}\end{pmatrix}\Gamma_{NN}^{-1}\begin{pmatrix}0\\ \Gamma_{0N}\\ \vdots\\ \Gamma_{N-1,N}\end{pmatrix}^* \tag{4.6}$$

Let us put

$$\Delta\Gamma = \begin{pmatrix} & & & 0\\ & \Gamma & & \vdots\\ 0 & \cdots & 0 & 0\end{pmatrix} - \begin{pmatrix} 0 & 0 & \cdots & 0\\ \vdots & & & \\ & & \Gamma & \\ 0 & & & \end{pmatrix}$$

Then

$$\Delta\Gamma = \begin{pmatrix} & & & & 0\\ & & \Gamma - Z\Gamma Z^* & & -\Gamma_{N0}\\ & & & & \vdots\\ & & & & -\Gamma_{NN-1}\\ 0 & -\Gamma_{0N} & \cdots & -\Gamma_{N-1N} & -\Gamma_{NN}\end{pmatrix} \tag{4.7}$$

Thus

$$\sum_{i,j=0}^{N+1}(\Delta\Gamma)_{ij}z^i(\omega^*)^j = \sum_{i,j=0}^{N}(\Gamma - Z\Gamma Z^*)_{ij}z^i(\omega^*)^j$$

$$- z^{N+1}\omega^*\left(\Gamma_{0N}+\cdots+(\omega^*)^{(N-1)}\Gamma_{N-1,N}\right)$$

$$- z^{N+1}(\omega^*)^{(N+1)}\Gamma_{NN} - (\omega^*)^{(N+1)}z\left(\Gamma_{N0}+\cdots+\Gamma_{N,N-1}z^{N-1}\right)$$

and using equation (4.6), we obtain

$$\begin{aligned}(1 - z\omega^*)G(z, \omega) &= Q_N(z)\Gamma_{00}^{-1}Q_N(\omega)^* \\ &- z\left(P_N(z) - z^N\Gamma_{NN}\right)\Gamma_{NN}^{-1}\left(P_N(\omega)^* - (\omega^*)^N\Gamma_{NN}\right)\omega^* \\ &- z^{N+1}\omega^*\left(P_N(\omega)^* - (\omega^*)^N\Gamma_{NN}\right) \\ &- z^{N+1}(\omega^*)^{(N+1)}\Gamma_{NN} - (\omega^*)^{N+1}z\left(P_N(z) - z^N\Gamma_{NN}\right)\end{aligned}$$

This leads to (4.1) after simple transformations. □

COROLLARY 4.1. *If the blocks* Γ_{00} *and* Γ_{NN} *of* T_N^{-1} *are invertible, they have the same signature; in particular* $\Gamma_{00} > 0$ *if and only if* $\Gamma_{NN} > 0$ *and* $\Gamma_{00} < 0$ *if and only if* $\Gamma_{NN} < 0$.

PROOF. It suffices to write

$$(1 - z\omega^*)G(z, \omega) = Q_N(z)\Gamma_{00}^{-1}Q_N(\omega)^* - z\omega^*P_N(z)\Gamma_{NN}^{-1}P_N(\omega)^*$$

and to take a point $z = \omega^*$ on the unit circle such that $\det Q_N(\omega) \neq 0$ and $\det P_N(\omega) \neq 0$. □

In general P_N and Q_N may have zeros on the unit circle but the following corollary holds

COROLLARY 4.2. *If* T_N *and* T_{N-1} *are invertible, then there does not exist a non-zero vector* c *such that, for some complex point* ω,

$$P_N(\omega)^*c = Q_N(\omega)^*c = 0$$

PROOF. If such a pair (ω, c) exists, formula (4.1) leads to

$$\sum z^i(\omega^*)^j\Gamma_{ij}c = 0 \qquad \forall z \in \mathbf{C}$$

i.e.,

$$(I_p, zI_p, \ldots, z^NI_p)\Gamma\begin{pmatrix} c \\ \omega c \\ \vdots \\ \omega^N c\end{pmatrix} = 0 \qquad \forall z \in \mathbf{C}$$

which contradicts the invertibility of Γ, unless $c = 0$. □

5. A GENERALIZATION OF THE THEOREM OF M.G. KREIN

In this section we obtain the following generalization of Theorem 3.1.

THEOREM 5.1. *Let T_N be the block Toeplitz matrix defined in (3.1) and suppose that both T_N and T_{N-1} are invertible. Let ν_{N-1} be the number of negative eigenvalues of T_{N-1} and suppose Γ_{00} is positive or negative definite. Then $\det P_N$ and $\det Q_N$ have no zeros on the unit circle. Moreover,*

a) *If $\Gamma_{00} > 0$, then $\det P_N$ has ν_{N-1} zeros outside $\mathbb{D}$ and $(Np - \nu_{N-1})$ zeros inside $\mathbb{D}$.*

b) *If $\Gamma_{00} < 0$, then $\det P_N$ has ν_{N-1} zeros inside $\mathbb{D}$ and $(Np - \nu_{N-1})$ zeros outside $\mathbb{D}$.*

PROOF. We will focus on the case $\Gamma_{00} > 0$. Equation (4.5) with $\begin{pmatrix} A & B \\ C & D \end{pmatrix} = T_N$ and $D = T_{N-1}$ implies that T_N and T_{N-1} have the same number of negative eigenvalues. We denote by p_N and q_N the polynomials

$$p_N(z) = zP_N(z)\Gamma_{00}^{-1/2}\ , \qquad q_N(z) = Q_N(z)\Gamma_{00}^{-1/2}$$

Formula (4.1) becomes

$$G(z,\omega) = \frac{q_N(z)q_N^*(\omega) - p_N(z)p_N^*(\omega)}{1 - z\omega^*} \tag{5.1}$$

The polynomials $\det p_N$ and $\det q_N$ (and hence $\det P_N$ and $\det Q_N$) have no zeros on the unit circle. Indeed, let ω with $|\omega| = 1$ be such that $p_N^*(\omega)c = 0$. Then equation (5.1) implies that $q_N^*(\omega)c = 0$, and hence, by Corollary 2.1, $c = 0$.

Let us set $u_N(z) = q_N^{-1}(z)p_N(z)$. The function u_N is unitary on the unit circle. The rest of the proof is based on properties of u_N and is divided into steps, to ease the presentation.

STEP 1. *The MacMillan degree of the function u_N is $(N+1)p$ and $\nu(u_N) = \nu_{N-1}$.*

PROOF OF STEP 1. By Theorem 2.3, $\deg u_N$ is equal to the dimension of the linear span of the functions $z \to k_{u_N}(z,\omega)c$, where ω varies in the points of analyticity of u_N and c varies in C_p. This span has the same dimension as the linear span of the functions $z \to G(z,\omega)c$, where ω varies in C and c varies in C_p. The latter linear span is equal to the space of polynomials of degree less than or equal to N with coefficients in C_p and hence has dimension $(N+1)p$. Indeed, any element in this linear span, $f(z) = \sum_1^m G(z,\omega_i)c_i$, can be written as

$$f(z) = y_0 + y_1 z + \cdots + y_N z^N$$

where

$$\begin{pmatrix} y_0 \\ \vdots \\ y_N \end{pmatrix} = T_N^{-1} \begin{pmatrix} \sum_1^m c_i \\ \vdots \\ \sum_1^m c_i(\omega_i^*)^N \end{pmatrix}$$

Conversely, any vector $\begin{pmatrix} y_0 \\ \vdots \\ y_N \end{pmatrix}$ of $C_{(N+1)p}$ can be written in such a form by fixing $m = N$ and choosing N points $\omega_1, \ldots, \omega_N$ such that $\omega_i \neq \omega_j$ for $i \neq j$. Thus, $\deg u_N = (N+1)p$. The number $\nu(u_N)$ is equal to the number of negative squares of the function $G(z, \omega)$. Let r be an integer, $\omega_1, \ldots, \omega_r$ points in C, and $c_1, \ldots, c_r$ in C_p. Then

$$\left(c_j^* G(\omega_j, \omega_i) c_i\right)_{i,j=1}^r = X^* \Gamma_N^{-1} X$$

with

$$X = \begin{pmatrix} c_1 & c_2 & \cdots & c_r \\ \omega_1 c_1 & \omega_2 c_2 & \cdots & \omega_r c_r \\ \vdots & \vdots & & \vdots \\ \omega_1^N c_1 & \omega_2^N c_2 & \cdots & \omega_r^N c_r \end{pmatrix}$$

Let us choose $r = (N+1)p$, all the ω_i different, and $[c_1, \ldots, c_r] = [I_p, \ldots, I_p]$. Then the matrix X is invertible. Indeed, let $y_0, \ldots, y_n$ be in C_p such that $(y_0^*, \ldots, y_N^*)X = 0$, or

$$(y_0^* c_1) + \omega_1(y_0^* c_1) + \cdots + \omega_1^N (y_0^* c_1) = 0$$

$$\vdots$$

$$(y_0^* c_r) + \omega_r(y_0^* c_r) + \cdots + (\omega_r^N)(y_0^* c_r) = 0$$

Taking into account that $c_{p+1} = c_1, c_{p+2} = c_2, \ldots,$ we can regroup these equations into p subsystems and show that all the $y_i^* c_j = 0$ for $i = 0, \ldots, N, j = 1, \ldots, r$. Hence, the function $G(z, \omega)$ has ν_{N-1} negative squares, since T_N has ν_{N-1} negative eigenvalues. This ends the proof of Step 1.

STEP 2. *The matrix polynomial p_N has zeros inside $\mathbb{D}$ of total multiplicity at least $(N+1)p - \nu_{N-1}$.*

PROOF OF STEP 2. By Theorem 2.7 the function u_N may be represented in the form

$$u_N(z) = b_1(z) b_2^{-1}(z) \tag{5.2}$$

where b_1 and b_2 are $C_{p\times p}$ valued Blaschke-Potapov products such that

$$\deg b_1 = (N+1)p - \nu_{N-1}$$

$$\deg b_2 = \nu_{N-1}$$

and equation (5.2) can be rewritten as

$$p_N b_2 = q_N b_1 \tag{5.3}$$

Let ω be a zero of $\det b_1$, of multiplicity r. Then ([BGK] p.45-46) there are k independent eigenvectors $x_{11}, \ldots, x_{k1}$, i.e., vectors such that $b_1(\omega)x_{i1} = 0, i = 1, \ldots, k$, with which are asssociated k Jordan chains of lengths $r_1, \ldots, r_k$, and $r = r_1 + \cdots + r_k$. For a $C_{p\times p}$ valued function a analytic at the point ω, let us denote by $L_\ell(a)$ the matrix

$$L_\ell(a) = \begin{pmatrix} a(\omega) & 0 & \cdots & 0 \\ a'(\omega) & a(\omega) & \cdots & 0 \\ a^{(2)}(\omega)/2 & a'(\omega) & \cdots & 0 \\ \vdots & \vdots & & \vdots \\ a^{(\ell)}(\omega)/\ell! & a^{(\ell-1)}(\omega)/_{(\ell-1)!} & \cdots & a(\omega) \end{pmatrix} \tag{5.4}$$

It is easy to check that $L_\ell(ab) = L_\ell(a)L_\ell(b)$ and thus (5.3) implies

$$L_\ell(p_N)L_\ell(b_2) = L_\ell(q_N)L_\ell(b_1) \tag{5.5}$$

Let $x_{i1}, \ldots, x_{ir_i}$ be a Jordan chain of length r_1 corresponding to the eigenvector x_{11}. Then

$$L_{r_1}(b_1)\begin{pmatrix} x_{i1} \\ \vdots \\ x_{ir_i} \end{pmatrix} = 0$$

By Corollary 2.1, the vector $y_{i1} = b_2(\omega)x_{i1}$ is non-zero and thus equation (5.5) implies that $y_{i1}, \ldots, y_{ir_i}$, defined by

$$L_{r_1}(b_2)\begin{pmatrix} x_{i1} \\ \vdots \\ x_{ir_i} \end{pmatrix} = \begin{pmatrix} y_{i1} \\ \vdots \\ y_{ir_i} \end{pmatrix}$$

form a Jordan chain of length r_1 for p_N at the point ω.

The k eigenvectors $y_{11}, \ldots, y_{k1}$ are linearly independent. Indeed if for some complex numbers $\alpha_1, \ldots, \alpha_k$,

$$\sum \alpha_i y_{i1} = 0$$

then the vector $c = \sum \alpha_i x_{i1}$ is such that $b_1(\omega)c = b_2(\omega)c = 0$ and thus is zero. By Corollary 2.1, this forces the α_i to be zero. Hence the multiplicity of $\det p_N$ at ω is at least $r = r_1 + r_2 + \cdots + r_k$.

The zeros of b_1 are all inside $\mathbb{D}$ and have total multiplicity $(N+1)p - \nu_N$, hence the zeros of p_N inside $\mathbb{D}$ have a total multiplicity at least equal to $(N+1)p - \nu_{N-1}$.

STEP 3. $\nu(v_N) = (N+1)p - \nu(u_N)$, *where* $v_N(z) = u_N(1/z)$.

PROOF OF STEP 3. The function u_N is unitary on the unit circle and satisfies $u_N(z)u_N^*(1/z^*) = I_p$. Hence, $v_N^*(z)u_N(z^*) = I_p$ and

$$\frac{I_p - v_N(z)v_n^*(\omega)}{1 - z\omega^*} = (u_N^*(z^*))^{-1}\frac{u_N^*(z^*)u_N(\omega^*) - I_p}{1 - z\omega^*}(u_N(\omega^*))^{-1} \tag{5.6}$$

By Theorem 2.3 the functions

$$\frac{I_p - u_N^*(z^*)u_N(\omega^*)}{1 - \omega^*} \quad \text{and} \quad \frac{I_p - u_N(\omega^*)u_N^*(z^*)}{1 - z\omega^*}$$

have the same number of negative squares. Replacing ω^* by z and z by ω^*, we see that this last function has $\nu(u_N)$ negative squares. From equation (5.6) it follows that $\nu(v_N) = (N+1)p - \nu(u_N)$.

Let us now conclude the proof of the theorem for the case $\Gamma_{00} > 0$.

By Theorem 2.3 there exist two $C_{p\times p}$ valued Blaschke-Potapov products such that

$$v_N = b_3 b_4^{-1}$$

with

$$\deg b_3 = \nu_{N-1}, \qquad \deg b_4 = (N+1)p - \nu_{N-1}$$

and hence,

$$z^{N+1}p_N(1/z)b_4(z) = z^{N+1}q_N(1/z)b_3(z)$$

By the same argument on multiplicities as above, the polynomial $z^{N+1}p_N(1/z)$ now has at least v_{N-1} zeros inside $\mathbb{D}$. The constant term of this polynomial is Γ_{NN} and thus is invertible. Hence, all the zeros of $z^{N+1}p_N(1/z)$ are different from $z = 0$, and so p_N has at least v_{N-1} zeros outside $\mathbb{D}$. Since $\deg p_N = (N+1)p$ we conclude that p_N has exactly v_{N-1} zeros outside $\mathbb{D}$ and exactly $(N+1)p - v_{N-1}$ inside $\mathbb{D}$. The theorem is proved for the case $\Gamma_{00} > 0$ since $p_N(z) = zP_N(z)\Gamma_{00}^{-1/2}$.

The case $\Gamma_{00} < 0$ is treated similarly. □

As a corollary of the proof of this theorem it follows that if $\Gamma_{00} > 0$, then the polynomial $\det Q_N$ has at least ν_{N-1} zeros inside $\mathbb{D}$ and the polynomial $\det z^N Q_N(\frac{1}{z})$

has at least $(Np - \nu_{N-1})$ zeros inside $\mathbb{D}$, while if $\Gamma_{00} < 0$, then $\det Q_N$ has at least $(Np - \nu_{N-1})$ zeros inside $\mathbb{D}$ and $\det z^N Q_N(\frac{1}{z})$ has at least ν_{N-1} zeros inside $\mathbb{D}$. In general, one cannot be more precise since Γ_{N0}, the non-constant term of $z^N Q_N(\frac{1}{z})$, may fail to be invertible.

REFERENCES

[AG1] D. Alpay and I. Gohberg, Unitary rational functions and orthogonal matrix polynomials, preprint, 1987.

[AG2] D. Alpay and I Gohberg, Unitary rational matrix functions, in press.

[BGK] H. Bart, I. Gohberg and M.A. Kaashoek, *Minimal factorizations of matrix and Operator Functions*, OT1: Operator Theory: Advances and Applications, Vol 1, Birkhäuser Verlag, Basel, 1979.

[EGL] R. Ellis, I. Gohberg and D. Lay, On two theorems of M.G. Krein concerning on the unit circle, *Integral Equation and Operator Theory* **11** (1988), 87-104.

[GLR] I. Gohberg, P. Lancaster and L. Rodman, *Matrices and indefinite scalar products*, OT8: Operator Theory: Advances and Applications Vol 8, Birkhäuser Verlag, Basel, 1983.

[K] M.G. Krein, Distribution of roots of polynomials orthogonal on the unit circle with respect to a sign alternating weight, *Theor. Funkcii Funkcional Anal.i. Prilozen.* **2** (1966), 131-137 (Russian).

[KL] M.G. Krein and H. Langer, Über die verallgemeinerten Resolventen und die charakteristische Funktion eines isometrischen Operators im Raume π_k, *Colloquia Mathematica Societatis Janos Bolyai* 5. *Hilbert space operators*, Tihany, (Hungary), 1970, 353-399.

[S] L.A. Sakhnovich, Factorization problems and operator identities, *Russian Mathematical Surveys* **41**:1, (1986), 1-64.

Daniel Alpay
Department of Electronic Systems
Tel-Aviv University
Tel-Aviv, 69978 Israel

and

Israel Gohberg
Raymond and Beverly Sackler
Faculty of Exact Sciences
School of Mathematical Sciences
Tel-Aviv University
Tel-Aviv, 69978 Israel

Present address:

Daniel Alpay
Department of Mathematics
Groningen University
P.O.B. 800, 9700AV
Groningen, Holland

Operator Theory:
Advances and Applications, Vol. 34

n-ORTHONORMAL OPERATOR POLYNOMIALS

Aharon Atzmon

1. INTRODUCTION

If μ is a positive Borel measure with infinite support on the unit circle T in the complex plane C, then the Gram-Schmidt process applied in $L^2(\mu)$ to the sequence of polynomials $1, z, z^2, \ldots,$ yields an orthonormal sequence of polynomials in $L^2(\mu)$. The $n+1$-th polynomial p in this sequence is uniquely determined up to a scalar multiple of modulus one by the conditions

$$\int_T p(\lambda)\overline{\lambda}^j d\mu(\lambda) = 0, \qquad j = 0, \ldots, n-1, \tag{1.1}$$

$$\int_T |p(\lambda)|^2 d\mu(\lambda) = 1 \tag{1.2}$$

and the requirement that p is of degree n. This fact and the paper [8] of M.G. Krein motivate the following

DEFINITION 1.1. A polynomial p of degree n is called *n-orthonormal* if there exists a signed (that is, real valued) Borel measure μ on T such that conditions (1.1) and (1.2) hold. Every such measure μ is called a *generating measure* for p.

It is known [7, p.43] that a polynomial p of degree n is n- orthonormal with respect to a positive generating measure, if and only if $p(\lambda) \neq 0$ for $|\lambda| \geq 1$, and in this case one of the generating measures μ is given by the formula

$$d\mu(e^{i\theta}) = \frac{1}{2\pi} \cdot \frac{d\theta}{|p(e^{i\theta})|^2}. \tag{1.3}$$

On the other hand in [8] M.G. Krein, announced with some hints on the proof, the following result:

A polynomial p of degree n is n-orthonormal if and only if

(1.4) $$|p(\lambda)| + |p(\overline{\lambda}^{-1})| > 0, \qquad \text{for every} \quad \lambda \neq 0 .$$

A complete proof of Krein's result is given in [4]. Neither [8] nor [4] contain formulas for the generating measures.

In this paper we give a short proof of an extension of Krein's result to matrix polynomials (Theorem 3.1) and obtain formulas for the generating measures in terms of certain objects associated with the n-orthonormal polynomial. These formulas are also new in the scalar case. Our result also implies (Corollary 3.2) one of the results in [6], which is also an extension (in somewhat different form) of Krein's Theorem to matrix polynomials.

We also obtain (Theorem 3.4) from our main result the characterization of n-orthonormal matrix polynomials which have positive generating matrix measure, along with an extension of formula (1.3) to this setting.

Our methods are completely different from those in [4] and [6]. Whereas these authors apply methods from matrix theory and results concerning the inversion of Toeplitz matrices, we use only elementary facts about Fourier series of operator functions.

In Section 2 we introduce our notations and definitions, and in Section 3 we state and prove our main results. We devote an Appendix to a direct proof of Krein's result for the scalar case, which is independent from the rest of the paper. Hence the reader interested only in the scalar case, may proceed directly to the Appendix.

I wish to express my thanks to Israel Gohberg from whom I learned about these problems and who encouraged me to write this paper, and to Leonid Lerer for valuable discussions on these topics.

2. NOTATIONS, DEFINITIONS AND PRELIMINARIES

In what follows H will denote a complex Hilbert space and $\mathcal{L}(H)$ the algebra of bounded linear operators on H. For a continuous function $f : T \to \mathcal{L}(H)$, we denote for very integer j, by $\widehat{f}(j)$ its j-th Fourier coefficient, that is, the element in $\mathcal{L}(H)$ defined by (the $\mathcal{L}(H)$-valued Rieman integral)

$$\widehat{f}(j) = \frac{1}{2\pi} \int_0^{2\pi} f(e^{i\theta}) e^{-ij\theta} d\theta \ .$$

We shall denote by W the Wiener algebra of $\mathcal{L}(H)$-valued functions, that is the algebra of all continuous functions $f : T \to \mathcal{L}(H)$ such that $\sum_{j=-\infty}^{\infty} \|\widehat{f}(j)\| < \infty$. The sub-algebra of all functions g in W such that $\widehat{g}(j) = 0$ for $j < 0$ (respectively, for $j \leq 0$) will be denoted by W_+ (respectively, by W_+^0). For every integer $n \geq 0$, we shall denote by W_n the vector space of all functions f in $\mathcal{L}(H)$ such that $\widehat{f}(j) = 0$ for $|j| > n$, that is, the set of all $\mathcal{L}(H)$-valued trigonometric polynomials of degree at most n. For a function f in W, we shall denote by f^* the function defined by $f^*(\lambda) = (f(\lambda))^*$, $\lambda \in T$. Since $\widehat{f^*}(j) = \widehat{f}(-j)^*$ for every integer j, f^* is also in W. If $f = f^*$ then f is called *hermitian.*

We shall denote by Π the algebra of polynomials in the complex variable z with coefficients in $\mathcal{L}(H)$, identified in the obvious way, with the sub-algebra $(\bigcup_{n=0}^{\infty} W_n) \cap W_+$ of W_+. For every integer $n \geq 0$, we shall denote by Π_n the vector space of polynomials in Π of degree not exceeding n, that is $\Pi_n = \Pi \cap W_n$. We denote by s the polynomial $s(z) = Iz$ in Π, where I is the identity of $\mathcal{L}(H)$. Thus every polynomial p in Π_n can be written in the form $p = \sum_{j=0}^{n} \widehat{p}(j)s^j$.

DEFINITION 2.1. A mapping $\Phi : \Pi \times \Pi \to \mathcal{L}(H)$ will be called an *operator valued hermitian form on* Π if it satisfies the following conditions:

(2.1) for every polynomial q in Π the mapping $p \to \Phi(p, q)$ of Π into $\mathcal{L}(H)$ is linear.

(2.2) $\Phi(Ap, q) = A\Phi(p, q)$, for every A in $\mathcal{L}(H)$ and p, q in Π.

(2.3) $\Phi(p, q)^* = \Phi(q, p)$, for all p, q in Π.

Note that conditions (2.2) and (2.3) imply that

(2.4) $\Phi(Ap, Bq) = A\Phi(p, q)B^*$, for all A, B in $\mathcal{L}(H)$ and p, q in Π.

If in addition the form Φ satisfies the condition

(2.5) $\Phi(sp, sq) = \Phi(p, q)$ for all p, q in Π,

then Φ is called a *stationary hermitian form.*

We shall denote by S the set of all stationary hermitian forms on Π.

A hermitian form Φ on Π is called *positive* if the operator $\Phi(p, p)$ is positive definite for every p in Π.

If f is a hermitian function in W, then it is easy to verify that the mapping $\Phi_f : \Pi \times \Pi \to \mathcal{L}(H)$ defined by

$$\Phi_f(p,q) = \frac{1}{2\pi}\int_0^{2\pi} p(e^{i\theta})f(e^{i\theta})q^*(e^{i\theta})d\theta\ , \quad p,q \in \Pi$$

is in S. We call Φ_f the *form associated with* f.

Forms of type Φ_f, as well as more general forms defined in terms of operator measures, appear in [3],[10], and [11].

In the scalar case $H = C$, also $\mathcal{L}(H) = C$, and every signed Borel measure μ on T, defines a form Φ_μ in S by the formula

$$\Phi_\mu(p,q) = \int_T p(\lambda)\overline{q(\lambda)}d\mu(\lambda), \qquad p,q \in \Pi\ .$$

We are now in a position to define the notion of n-orthonormal operator polynomials.

DEFINITION 2.2. Let Φ be a form in S and let n be a positive integer. A polynomial p in Π_n is called *n-orthonormal with respect to* Φ and Φ is called a *generating form for* p if the following conditions hold:

(2.6) $$\Phi(p,s^j) = 0, \qquad j = 0,1,\ldots,n-1\ .$$

(2.7) $$\Phi(p,p) = I\ .$$

Notice that (2.6) is equivalent to the condition

(2.8) $$\Phi(p,q) = 0, \qquad \text{for every } q \text{ in } \Pi_{n-1}\ .$$

DEFINITION 2.3. A polynomial p in Π_n is called *n-orthonormal* if p is n-orthonormal with respect to some form in S.

In the sequel we shall denote for every positive integer n, by O_n the set of all n-orthonormal polynomials, and for every form Φ in S we denote by $O_n\Phi$ the set of all n-orthonormal polynomials with respect to Φ. The set of all n-orthonormal polynomials which have a positive generating form will be denoted by O_n^+.

In order to establish the connection between the results of this paper and some of the results in [6] we shall need to consider in the sequel also a more general class of polynomials. For every positive integer n and every operator a in $\mathcal{L}(H)$ we shall denote by $O_n(a)$ the set of all polynomials p in Π_n for which there exists a form Φ in S such that (2.6) holds and

(2.9) $$\Phi(p,p) = a\ .$$

It follows from (2.3) that if (2.9) holds for some p in Π and Φ in S then a is self-adjoint.

It is clear that in the scalar case, an n-orthonormal polynomial in the sense of Definition 1.1 is also n-orthonormal in the sense of Definition 2.3. We shall see that the converse is also true, that is, in the scalar case both definitions are equivalent. For this we first have to define the Fourier coefficients of a form in S.

DEFINITION 2.4. If Φ is a form in S then for every integer j, the j-th Fourier coefficient $\widehat{\Phi}(j)$ of Φ is the element in $\mathcal{L}(H)$ defined by $\Phi(I, s^j)$ for $j \geq 0$ and by $\Phi(s^{-j}, I)$ for $j < 0$.

It is easy to see that if f is a hermitian function in W then $\widehat{\Phi}_f(j) = \widehat{f}(j)$ for every integer j, and that in the scalar case, if μ is a signed Borel measure on T, then for every integer j, $\widehat{\Phi}_\mu(j) = \widehat{\mu}(j)$ (the j-th Fourier Stieltjes coefficient of μ).

For the description of Fourier coefficients of forms in S, it is convenient to introduce the following:

DEFINITION 2.5. A sequence $(A_j)_{j=-\infty}^{\infty}$ in $\mathcal{L}(H)$ is called a *hermitian sequence* if $A_j^* = A_{-j}$ for every integer j.

It is easily verified that for every form Φ in S, the sequence $(\widehat{\Phi}(j))_{j=-\infty}^{\infty}$ is hermitian, and condition (2.5) implies that $\Phi(s^j, s^k) = \widehat{\Phi}(k-j)$ for every pair of integers $j \geq 0$ and $k \geq 0$, that is, the infinite matrix $(\Phi(s^j, s^k))_{j,k=0}^{\infty}$ is a self adjoint Toeplitz matrix with entries in $\mathcal{L}(H)$.

From these observations it follows that if Φ is a form in S, then for all p and q in Π_n we have the equality

$$\Phi(p,q) = \sum_{j,k=0}^{n} \widehat{p}(j)\widehat{\Phi}(k-j)\widehat{q}(k)^* \ . \tag{2.10}$$

This shows that a form Φ in S is uniquely determined by its Fourier coefficients, and moreover, that the restriction of Φ to $\Pi_n \times \Pi_n$ is uniquely determined already by the coefficients $\widehat{\Phi}(0), \widehat{\Phi}(1), \ldots, \widehat{\Phi}(n)$.

It is easily verified that if $(A_j)_{j=-\infty}^{\infty}$ is a hermitian sequence in $\mathcal{L}(H)$, then the mapping $\Psi : \Pi \times \Pi \to \mathcal{L}(H)$ defined by

$$\Psi(p,q) = \sum_{j,k=0}^{\infty} \widehat{p}(j) A_{k-j} \widehat{q}(k)^*, \qquad p, q \in \Pi \ ,$$

is in S, and $\widehat{\psi}(j) = A_j$ for every integer j. Consequently, the mapping $\Phi \to (\widehat{\Phi}(j))_{j=-\infty}^{\infty}$ is a one to one correspondence between S and the set of all hermitian sequences in $\mathcal{L}(H)$, or equivalently, the mapping $\Phi \to (\widehat{\Phi}(k-j))_{k,j=0}^{\infty}$, is a one to one correspondence between S and the set of all self-adjoint infinite Toeplitz matrices with entries in $\mathcal{L}(H)$.

It also follows from these observations that for every form Φ in S, the function f in W_n defined by $f(\lambda) = \sum_{j=-n}^{n} \widehat{\Phi}(j)\lambda^j, \lambda \in T$, is hermitian and the restrictions of the forms Φ and Φ_f to $\Pi_n \times \Pi_n$ coincide. Hence if p is a polynomial in Π_n then p is in $O_n\Phi(a)$ if and only if p is in $O_n\Phi_f(a)$. Thus we have the following:

PROPOSITION 2.1. *A polynomial p in Π_n is in $O_n(a)$ if and only if p is in $O_n\Phi_f(a)$ for some hermitian function f in W_n.*

That is, every polynomial in $O_n(a)$ has a generating form which is associated with some hermitian trigonometric polynomial with values in $\mathcal{L}(H)$ of degree not exceeding n. It follows from this observation that in the scalar case, Definitions 1.1 and 2.3 are equivalent.

For polynomials p in Π_n such that $\widehat{p}(n)$ is invertible, the condition of belonging to $O_n(a)$ can also be expressed in terms of a matrix equation. To see this, assume that p is a polynomial in $O_n\Phi(a)$. It follows from (2.3) and (2.10) that (2.6) and (2.9) are equivalent to the equations

$$\sum_{k=0}^{n} \widehat{\Phi}(k-j)\widehat{p}(k)^* = 0, \qquad j = 0,\ldots,n-1\ , \tag{2.11}$$

$$\widehat{p}(n)\sum_{k=0}^{n} \widehat{\Phi}(k-n)\widehat{p}(k)^* = a\ . \tag{2.12}$$

Equation (2.12) implies that if a is invertible then $\widehat{p}(n)$ is right invertible, hence invertible if H is finite dimensional.

Consider the $(n+1)\times(n+1)$ Toeplitz matrix Γ with entries in $\mathcal{L}(H)$, defined by $\Gamma = (\widehat{\Phi}(j-k))_{j,k=0}^{n}$. Since the sequence $(\widehat{\Phi}(j))_{j=-\infty}^{\infty}$ is hermitian, the matrix Γ is self-adjoint. If $\widehat{p}(n)$ is invertible, then equations (2.11) and (2.12) are equivalent to the matrix equation

$$\Gamma \begin{pmatrix} \widehat{p}(n)^* \\ \vdots \\ \widehat{p}(0)^* \end{pmatrix} = \begin{pmatrix} \widehat{p}(n)^{-1}a \\ 0 \\ \vdots \\ 0 \end{pmatrix} \tag{2.13}$$

Conversely, if p is a polynomial in Π_n such that $\widehat{p}(n)$ is invertible, and there exists an $(n+1)\times(n+1)$ self-adjoint Toeplitz matrix $\Gamma = (t_{j-k})_{j,k=0}^{n}$ with entries in $\mathcal{L}(H)$, such that equation (2.13) holds, then equations (2.11) and (2.12) are satisfied with $\Phi = \Phi_f$, where f is the hermitian function in W_n defined by $f(\lambda) = \sum_{j=-n}^{n} t_j\lambda^j, \lambda \in T$, and therefore

p is in $O_n\Phi_f(a)$. Consequently, equation (2.13) characterizes the set of polynomials p in Π_n with $\widehat{p}(n)$ invertible, which are in $O_n(a)$.

We conclude this section by describing the connection between the classes $O_n(a)$ and a class of column vectors with entries in $\mathcal{L}(H)$ considered by the authors in [6]. These authors consider in [6, Sec. 7] the class of column vectors $x = \operatorname{col}(x_j)_{j=0}^n$ with $x_j \in \mathcal{L}(H)$, $j = 0, 1, \ldots, n$ and x_0 invertible, for which there exists an $(n+1) \times (n+1)$ self-adjoint Toeplitz matrix Γ with entries in $\mathcal{L}(H)$ such that the equation

$$\Gamma \begin{pmatrix} x_0 \\ \vdots \\ x_n \end{pmatrix} = \begin{pmatrix} I \\ 0 \\ \vdots \\ 0 \end{pmatrix} \tag{2.14}$$

holds. Let K_n denote this class of column vectors.

From equations (2.13) and (2.14) and the remark after (2.9) we deduce:

PROPOSITION 2.2. *Let $x = \operatorname{col}(x_j)_{j=0}^n$ be a column vector with $x_j \in \mathcal{L}(H), j = 0, 1, \ldots, n$ and assume that x_0 is invertible. Then x is in K_n if and only if the polynomial $p = \sum_{j=0}^{n} x_{n-j}^* s^j$ is in $O_n(x_0)$, and in this case x_0 is self-adjoint.*

3. CHARACTERIZATION OF O_n FOR FINITE DIMENSIONAL SPACES

Throughout this section we assume that the Hilbert space H is finite dimensional. To state our main result we need several notations.

In what follows n will denote a fixed positive integer and a an invertible operator in $\mathcal{L}(H)$. For a polynomial p in Π_n we shall denote by $\widetilde{p}$ the polynomial $\sum_{j=0}^{n} \widehat{p}(n-j)^* s^j$, that is $\widetilde{p}(\lambda) = \lambda^n p(\overline{\lambda}^{-1})^*$, for $\lambda \neq 0$. For every $p \in \Pi$ and every ordered pair $(u, v) \in W_+ \times W_+$, we denote

$$L(p, u, v) = pu + v\widetilde{p}\,.$$

For a polynomial p in Π , we introduce the sets:

$$E(p, a) = \{(u, v) \in W_+^0 \times W_+ : L(p, u, v) = a\}$$

and

$$E_n(p, a) = E(p, a) \cap (\Pi_n \times \Pi_n)$$

For a polynomial p in $O_n(a)$ we denote

$$F(p,a) = \{f \in W : f = f^*, p \in O_n \Phi_f(a)\}$$

and

$$F_n(p,a) = F(p,a) \cap W_n$$

For $a = I$ we shall denote these sets simply by $E(p), E_n(p), F(p)$ and $F_n(p)$.

Finally for every function g in W we denote by $R(g)$ and $Q(g)$ the functions in W_+ and W_+^0 respectively defined by

$$R(g) = \sum_{j=0}^{\infty} \hat{g}(j) e^{ij\theta}; \qquad Q(g) = -\sum_{j=1}^{\infty} \hat{g}(-j)^* e^{ij\theta} .$$

Note that with these notations, we have that $g = R(g) - Q(g)^*$ for every g in W.

We are now ready to state our main result.

THEOREM 3.1. *For a polynomial p in Π_n the following conditions are equivalent:*
(a) p is in $O_n(a)$.
(b) $E_n(p,a)$ is not empty.
(c) $E(p,a)$ is not empty.
Furthermore, if the polynomial p is in $O_n(a)$ then it has an inverse p^{-1} in W and the formula

$$f = p^{-1}(s^n v - u^*), \qquad (u,v) \in E(p,a) , \tag{3.1}$$

establishes a one to one mapping of $E(p,a)$ onto $F(p,a)$, and its inverse is given by the formulas

$$u = Q(pf); \qquad v = R(s^{*n} pf), \qquad f \in F(p,a) . \tag{3.2}$$

Finally in the scalar case $H = C$, formula (3.1) can also be written in the form

$$f = vv^* - uu^* , \quad (u,v) \in E(p,a) . \tag{3.3}$$

Before turning to the proof of the theorem we show that it implies the Theorem of Krein described in Section 1, and also one of the characterizations of the class K_n given in [6].

To show the first fact, consider a polynomial p of degree n, with scalar coefficients. By the well known Euclid algorithm for the greatest common divisor of a pair of polynomials (cf [1], Ch. 15), the polynomial p satisfies condition (b) of Theorem 3.1, if and only if, the polynomials sp and $\widetilde{p}$ have no common roots, and since $\widetilde{p}(0) = \overline{\widehat{p}(n)} \neq 0$, (b) is also equivalent to the condition that p and $\widetilde{p}$ have no common roots. Therefore by the definition of $\widetilde{p}$, we obtain that condition (b) of Theorem 3.1 is equivalent to condition (1.4), hence to the Theorem of Krein.

The second implication mentioned, follows from Proposition 2.2. It is easily verified that in view of the connection between the classes K_n and $O_n(a)$ given by that proposition, part (b) of Theorem 3.1. implies the following result from [6]:

COROLLARY 3.2. [6, Th. 7.2(a'), (b')]. *Let $x = \mathrm{col}(x_j)_{j=0}^{n}$ be a column vector with $x_j \in \mathcal{L}(H), j = 0, 1, \ldots, n$ and with x_0 invertible, and consider the polynomial $p = \sum_{j=0}^{n} x_{n-j}^* s^j$ in Π_n. Then x is in K_n if and only if x_0 is self-adjoint and there exist polynomials φ and ψ in Π_n such that*

$$p\varphi + \psi\widetilde{p} = x_0$$

We note that using the fact that x_0 is invertible a standard argument shows that one can impose on the polynomials ϕ and ψ in the above equation, also one of the additional conditions $\psi \in \Pi_{n-1}$ or $\varphi(0) = 0$.

We present the proof of Theorem 3.1 for $a = I$. The proof of the general case is exactly the same, and is obtained by replacing everywhere in the following proof, I by a.

One of the main steps towards the proof of Theorem 3.1 is contained in the following simple result:

LEMMA 3.3. *Let p be a polynomial in Π_n and let f be a hermitian element in W_n. Then p is in $O_n\Phi_f$ if and only if there exist functions u in W_+^0 and v in W_+ such that*

$$pf = s^n v - u^* \tag{3.4}$$

and

$$\widehat{v}(0)\widehat{p}(n)^* = I\ . \tag{3.5}$$

Furthermore if (3.4) holds then the functions u and v are given by the formulas

$$u = Q(pf); \qquad v = R(s^{*n}pf)\ . \tag{3.6}$$

Moreover, if f is in W_n then the functions u and v in (3.4) are in Π_n.

PROOF. Set $g = pf$. Then g is in W, and it is clear that p is in $O_n\Phi_f$ if and only if

$$(3.7) \qquad \widehat{g}(j) = 0, \qquad j = 0, 1, \ldots, n-1 ,$$

and

$$(3.8) \qquad \frac{1}{2\pi}\int_0^{2\pi} g(e^{i\theta})p^*(e^{i\theta})d\theta = I .$$

In view of (3.7), condition (3.8) is equivalent to

$$(3.9) \qquad \widehat{g}(n)\widehat{p}(n)^* = I .$$

It is easily seen that conditions (3.7) and (3.9) are equivalent to the existence of functions u in W_+^0 and v in W_+ such that conditions (3.4) and (3.5) hold. It is clear from the assumptions on u and v, that if (3.4) holds, then these functions are given by formulas (3.6). This also implies the last assertion, and the proof is complete.

PROOF OF THEOREM 3.1. For $a = I$.

(a)⇒(b): Assume that p is in O_n. Then by Proposition 2.1, there exists a hermitian function f in W_n such that p is in $O_n\Phi_f$. Thus by Lemma 3.3, there exist functions u and v in Π_n such that $\widehat{u}(0) = 0$, and such that (3.4) holds. We shall show that the pair (u, v) is in $E_n(p)$. Multiplying both sides of equation (3.4) from the right by p^*, we obtain that

$$(3.10) \qquad pfp^* = v\widetilde{p} - (pu)^*$$

Since f is hermitian, the left hand side of (3.10) is also hermitian, hence the same is true for the right hand side of this equation. But this is equivalent to the equality $L(p,u,v) = L(p,u,v)^*$. Thus $L(p,u,v)$ is a hermitian element in W_+ and therefore it is constant, so we have that

$$(3.11) \qquad L(p,u,v) = L(p,u,v)(0)$$

Taking into account the fact that $u(0) = \widehat{u}(0) = 0$ and equation (3.5), we obtain that $L(p,u,v)(0) = I$, and therefore (3.11) implies that (u,v) is in $E_n(p)$, hence (b) holds.

(b)⇒(c): This is clear since $E_n(p) \subset E(p)$.

(c)⇒(a): Assume that $E(p)$ contains the pair (u,v). We shall show that p has an inverse in W, and that the function f associated with p and (u,v) by formula (3.1) is in $F(p)$. The equality $L(p,u,v) = I$ implies that for every $\lambda \in T$ and every vector x in H, we have that

$$\|x\|^2 = (u(\lambda)x, p^*(\lambda)x) + \lambda^n(p^*(\lambda)x, v(\lambda)x) .$$

This shows that $p^*(\lambda)$ is one to one, and since H is finite dimensional, it follows that $p(\lambda)$ is invertible. Thus $p(\lambda)$ is invertible for every $\lambda \in T$, hence by continuity, also for every λ in some open annulus which contains T. Therefore the operator function $p^{-1}(\lambda)$ is analytic in this annulus, hence its restriction to T is in W. Consequently p has an inverse p^{-1} in W. Let f be the function given by formula (3.1). Then f is in W, and equation (3.4) also holds. Using the fact that $L(p,u,v) = I$, and remembering that $\widehat{u}(0) = 0$, we obtain that p and v satisfy (3.5) and that $L(p,u,v)$ is hermitian. The last fact is equivalent to the identity

$$p(s^{*n}v - u) = (s^n v - u^*)p^*$$

which by (3.1) implies that f is hermitian. Thus by Lemma 3.3, p is in $O_n\Phi_f$, and the implication is proved.

To complete the proof of the theorem it remains to show that in the scalar case, (3.1) implies (3.3). To show this, assume that $H = C$ and that f is given by formula (3.1), where the pair (u,v) is in $E(p)$. Using the commutativity of products in this case, we obtain form (3.1) and the identity $L(p,u,v) = I$ the identities

$$pf = s^n v - u^*$$

$$\widetilde{p}f = v^* - s^n u$$

and

$$f = pfu + v\widetilde{p}f .$$

Substituting the expressions for pf and $\widetilde{p}f$ from the first two identities into the right hand side of the last identity, we obtain formula (3.3). This concludes the proof of the theorem.

REMARKS

1. If p is in O_n and (u,v) is in $E_n(p)$ then it follows from formula (3.3) that in the scalar case, the function f defined by formula (3.1) is in W_n. This is not true in the general case.

2. If p is in O_n, then one can show by using Theorem 3.1 and [1, Ch.15, Th.4] that in the scalar case, $F_n(p)$ contains a single element. This is also not true in the general case.

We now turn to the characterization of the class O_n^+. This is given by:

THEOREM 3.4. *Let p be a polynomial in Π_n. Then p is in O_n^+ if and only if $p(\lambda)$ is invertible for every $|\lambda| \geq 1$ and $\widehat{p}(n)$ is invertible. Furthermore if these conditions are satisfied then the function $f = (p^*p)^{-1}$ is in $F(p)$.*

PROOF. Assume first that p is in O_n^+ and that Φ is a corresponding positive generating form. Let λ be a complex number such that $p(\lambda)$ is not invertible. We shall show that $|\lambda| < 1$. The proof of this fact is similar to that given in [9, Prop.1] for the scalar case. We write p in the form $p = p(\lambda) + (s - \lambda I)q$, where q is in Π_{n-1}. Using the assumptions on Φ, in particular (2.5), (2.7) and (2.8), we obtain from the identity

$$p + \lambda q = p(\lambda) + sq$$

that

$$I + (|\lambda|^2 - 1)\Phi(q,q) = p(\lambda)\widehat{\Phi}(0)p^*(\lambda) + p(\lambda)\Phi(I, sq) + \Phi(sq, I)p^*(\lambda) \tag{3.12}$$

Since $p(\lambda)$ is not invertible and H is finite dimensional, there exists a unit vector x in H such that $p^*(\lambda)x = 0$. Combining this with (3.12) we obtain that

$$(1 - |\lambda|^2)(\Phi(q,q)x, x) = 1$$

and therefore since the operator $\Phi(q,q)$ is positive definite we conclude that $|\lambda| < 1$. Since H is finite dimensional, it follows from (2.12) that $\widehat{p}(n)$ is invertible.

To prove the other direction, assume that $p(\lambda)$ is invertible for every $|\lambda| \geq 1$ and that $\widehat{p}(n)$ is invertible. By continuity, there exists a number $0 < r < 1$ such that $p(\lambda)$ is invertible for every $|\lambda| \geq r$, and therefore $\widetilde{p}$ is invertible for $0 < |\lambda| < r^{-1}$. Since $\widehat{p}(n)$ is invertible and $\widetilde{p}(0) = \widehat{p}(n)^*$, we obtain that $\widetilde{p}(\lambda)$ is invertible in the disc $|\lambda| < r^{-1}$. Therefore the operator function $\widetilde{p}^{-1}(\lambda)$ is analytic in the disc $|\lambda| < r^{-1}$, hence its restriction to T is in W_+. This implies that the ordered pair $(0, \widetilde{p}^{-1})$ is in $E(p)$, and consequently by Theorem 3.1, the function $f = s^n p^{-1}\widetilde{p}^{-1}$ is in $F(p)$. Noticing that $\widetilde{p}(\lambda) = \lambda^n p^*(\lambda)$ for $\lambda \in T$, we obtain that $f = (p^*p)^{-1}$ and therefore the generating form Φ_f is positive. This completes the proof of the theorem.

REMARK. Theorem 3.4 is equivalent to the relevant parts of Theorems 14 and 15 in [3] where the forms considered are defined in terms of positive operator measures on

T. The equivalence can be established by showing that if Φ is a positive form in S then the sequence $(\widehat{\Phi}(j))_{j=-\infty}^{\infty}$ is a positive definite operator function on the integer group Z, in the sense of [5, Sec. 8], and therefore by [5, Sec. 8, Th.7] or [3, Th.1] Φ is given by a positive operator measure on T. We omit the details.

We conclude this section with some comments on Theorem 3.1 for infinite dimensional spaces. The assumption that H is finite dimensional was used in Theorem 3.1, only during the proof of the implication (c)⇒(a), where this fact was used to show that p has an inverse in W. For the infinite dimensional case, one can prove by using the results of [2] that condition (c) implies that p has a right inverse in W. One can use this fact to define the function f in W by formula (3.1) where p^{-1} is replaced by a right inverse of p in W. If one can prove that this function is hermitian, then an application of Lemma 3.3 (which remains true in the infinite dimensional case) shows that condition (a) is satisfied. However, the argument used in the proof of Theorem 3.1 to show that f is hermitian, was based on the fact that p has an inverse in W. We do not know whether implication (c)⇒(a) of Theorem 3.1 is true in the infinite dimensional case.

APPENDIX: THE SCALAR CASE

In this section we give a proof of Krein's Theorem which is independent of Sections 2 and 3 and establish a formula for the generating measure. We begin with some notations and comments.

For every integer $k \geq 0$, we denote by Π_k (respectively, by W_k) the vector space of (scalar) polynomials (respectively, trigonometric polynomials) of degree at most k. We identify Π_k with the subspace of W_k consisting of all functions f in this space such that $\widehat{f}(j) = 0$, for $j < 0$. For a complex number z, we shall denote by z^* its complex conjugate.

If p is a polynomial in Π_n, then it is clear that it satisfies conditions (1.1) and (1.2) for some signed Borel measure μ on T, if and only if it satisfies these conditions for the measure $\frac{1}{2\pi} f d\theta$, where f is the function in W_n given by $f(\lambda) = \sum_{j=-n}^{n} \widehat{\mu}(j)\lambda^j, \lambda \in T$. Since μ is real valued, $\widehat{\mu}(j)^* = \widehat{\mu}(-j)$ for every integer j, and therefore f is also real valued. Consequently, a polynomial p of degree n is n-orthonormal, if and only if there exists a real valued trigonometric polynomial f in W_n such that the conditions

$$\text{(A.1)} \qquad \frac{1}{2\pi} \int_0^{2\pi} p(e^{i\theta})e^{-ij\theta} f(e^{i\theta})d\theta = 0, \qquad j = 0, 1, \ldots, n-1,$$

and

(A.2) $$\frac{1}{2\pi}\int_0^{2\pi} |p(e^{i\theta})|^2 f(e^{i\theta})d\theta = 1$$

hold.

For a polynomial p in Π_n we shall denote by $\widetilde{p}$ the polynomial defined by $\widetilde{p}(z) = z^n(p(z^{*-1}))^*$, for $z \neq 0$. If p is a polynomial of degree n, then $\widetilde{p}(0) = \widehat{p}(n)^* \neq 0$, and therefore (1.4) is equivalent to the condition that the polynomials $zp(z)$ and $\widetilde{p}(z)$ have no common roots, and therefore by the well known characterization of relatively prime polynomials (c.f. [1,p.194]) this is equivalent to the existence of polynomials r in Π_{n-1} and v in Π_n such that

(A.3) $$zr(z)p(z) + v(z)\widetilde{p}(z) = 1, \qquad \forall z \in C \ .$$

Also by [1, p.194, Th.4] the polynomials r and v are uniquely determined by the above conditions. Setting $u(z) = zr(z)$, we conclude that condition (1.4) is equivalent to the existence of polynomials u and v in Π_n, with $u(0) = 0$, such that

(A.4) $$pu + \widetilde{p}v = 1 \ .$$

By the above remarks, these polynomials u and v are uniquely determined.

In view of these observations, the Theorem of M.G. Krein in [8] follows from the following:

THEOREM A.1. *Let p be a polynomial of degree n. Then there exists a real valued trigonometric polynomial f in W_n such that conditions (A.1) and (A.2) hold, if and only if, there exist polynomials u and v in Π_n, with $u(0) = 0$, such that equation (A.4) is satisfied. Moreover, in this case the function f is uniquely determined by the formula*

(A.5) $$f = vv^* - uu^*$$

PROOF. Assume first that there exists a real valued function f in W_n such that conditions (A.1) and (A.2) hold. Set $g = pf$. Then g is in W_{2n} and conditions (A.1) and (A.2) are equivalent to the conditions

(A.6) $$\widehat{g}(j) = 0, \qquad j = 0, 1, \ldots, n-1 \ ,$$

and

(A.7) $$\widehat{g}(n)\widetilde{p}(0) = 1$$

It is clear that these two conditions are equivalent to the existence of polynomials u and v in Π_n such that

$$u(0) = 0 \tag{A.8}$$

$$v(0)\widetilde{p}(0) = 1 \tag{A.9}$$

and

$$g = e^{in\theta}v - u^* . \tag{A.10}$$

Explicitely, the polynomials u and v are given by $u(z) = -\sum_{j=1}^{n} \widehat{g}(-j)^* z^n$ and $v(z) = \sum_{j=0}^{n} \widehat{g}(j+n)z^j$. Consider the function $h = pfp^*$ in W_{3n}. It follows from (A.10) that

$$h = \widetilde{p}v - (pu)^* \tag{A.11}$$

Since f is real valued, the same is true for h and therefore by (A.11) we have that

$$\widetilde{p}v - (pu)^* = (\widetilde{p}v)^* - pu$$

which is equivalent to the identity (on T)

$$pu + \widetilde{p}v = (pu + \widetilde{p}v)^* \tag{A.12}$$

Thus the polynomial $q = pu + \widetilde{p}v$, which is in Π_{2n} is real valued on T, and therefore for every positive integer j, $\widehat{q}(j) = \widehat{q}(-j)^* = 0$. Hence q is constant, so that $q = q(0)$. But by (A.8) and (A.9), $q(0) = 1$, and consequently the polynomials u and v satisfy equation (A.4).

To prove the other direction, assume that there exist polynomials u and v in Π_n such that equations (A.4) and (A.8) hold. Let f be the function defined in terms of u and v by formula (A.5). It is clear that f is a real valued function in W_n. We shall show that f satisfies conditions (A.1) and (A.2). Again set $g = pf$. Taking complex conjugates of both sides of (A.4) and multiplying by $e^{in\theta}$, we obtain the identity

$$pv^* + \widetilde{p}u^* = e^{in\theta} \tag{A.13}$$

From (A.4) and (A.13) we obtain that

$$p(vv^* - uu^*) = e^{in\theta}v - u^*$$

that is, g satisfies (A.10). From (A.4) and (A.8) it follows that (A.9) also holds, and consequently g satisfies equations (A.6) and (A.7) which are equivalent to equations (A.1) and (A.2).

Finally to prove the uniqueness assertion, assume that f is a real valued function in W_n such that conditions (A.1) and (A.2) hold, and let u and v be the polynomials in Π_n which satisfy equations (A.4) and (A.8). Multiplying both sides of (A.4) by f we obtain the identity

$$f = pfu + \tilde{p}fv$$

and using (A.10) and the fact that f is real valued, we obtain the identities

$$pf = e^{in\theta}v - u^*$$

and

$$\tilde{p}f = v^* - e^{in\theta}u$$

Substituting the expressions given for pf and $\tilde{p}f$ by the last two identities into the right hand side of the first identity, we obtain formula (A.5). Thus f is uniquely determined by the polynomials u and v. Since as remarked before these polynomials are uniquely determined by equation (A.4) and condition (A.8), we conclude that f is uniquely determined by conditions (A.1) and (A.2). This completes the proof of the theorem.

Note that equation (A.3) and formula (A.5) provide an effective way for computing the real valued function f in W_n which satisfies conditions (A.1) and (A.2) for a given n-orthonormal polynomial p. The solutions r and v of equation (A.3) can be obtained by means of formula (5) in [1,p.193], and setting $u(z) = zr(z)$, the function f is computed from formula (A.5).

EXAMPLE. Applying the process described above to the polynomial $p(z) = 2z^3 + 1$, we obtain that the solutions of equation (A.3) are given by $r(z) = \frac{1}{6}z^2$ and $v(z) = -\frac{1}{3}z^3 + \frac{1}{2}$. Thus $u(z) = \frac{1}{6}z^3$ and formula (A.5) yields that $f(e^{i\theta}) = -\frac{1}{3}(\cos 3\theta - 1)$.

REFERENCES

1. Bôcher, M., Introduction to higher algebra, Macmillan, New York, 1957.
2. Bochner, S., and Philips, R.S., Absolutely convergent Fourier expansions for non commutative normed rings, Annals of Math., 43(1942), 409-418.
3. Delsarte, P., Genin, Y.V., and Kamp, Y.G., Orthogonal polynomial matrices on the unit circle, IEES Trans. Circuits Syst. Vol. CAS-25, No.3(1978), 149-160.
4. Ellis, R.L., Gohberg, I., and Lay, D.C., On two theorems of M.G. Krein concerning polynomials orthogonal on the unit circle, Integral Equations Operator Theory, 11(1988), 87-104.
5. Filmore, P.A., Notes on operator theory, Van Nostrand, New York, 1970.
6. Gohberg, I., and Lerer, L., Matrix generalizations of M.G. Krein theorems on orthogonal polynomials, this issue.
7. Grenander, O., and Szegö, G., Toeplitz forms and their applications, Univ. of Calif. Press, Berkeley, 1957.
8. Krein, M.G., On the distribution of roots of polynomials which are orthogonal on the unit circle with respect to an alternating weight, Theor. Funkciĭ Funkcional Anal. i. Priložen. Resp. Sb., Nr. 2(1966), 131-137. (Russian).
9. Landau, H.J., Maximum entropy and the moment problem, Bull. A.M.S.(1987), 47-77.
10. Rosenberg, M., The square integrability of matrix valued functions with respect to a non-negative hermitian measure, Duke Math. J., 31(1964), 291-298.
11. Wiener, N., and Masani, P., The prediction theory of multivariate stochastic processes, Part I, Acta Math., 98(1957), 111-150; Part II, Acta Math., 99(1959), 93-137.

School of Mathematical Sciences
Raymond and Beverly Sackler Faculty of Exact Sciences
Tel-Aviv University, Tel-Aviv 69978, Israel.

Operator Theory:
Advances and Applications, Vol. 34

EXTENSION OF A THEOREM OF M. G. KREIN ON ORTHO GONAL POLYNOMIALS FOR THE NONSTATIONARY CASE

A. BEN-ARTZI and I. GOHBERG

The theorem of Krein, concerning the location of the zeros of orthogonal polynomials in an indefinite metric, is extended to the nonstationary block case. The proof relies heavily on results concerning nonstationary Stein equations and dichotomy from the authors' paper [2] and [3].

1. INTRODUCTION

In this paper we prove a nonstationary block version of the following well known theorem of M.G. Krein.

Let $R = (R_{i-j})_{ij=0}^{m}$ be an invertible self-adjoint Toeplitz matrix and assume that $\det(R_{i-j})_{ij=0}^{k} \neq 0$ $(k = 0,1,\ldots,m-1)$. Denote $d_k = \det(R_{i-j})_{ij=0}^{k}$ and let P be the number of permanences and V be the number of variations of sign in the sequence $1, d_0, d_1, \ldots, d_{m-1}$. Let $p(\varsigma) = x_0\varsigma^m + \cdots + x_m$ be the polynomial whose coefficients satisfy

$$R\begin{pmatrix} x_0 \\ x_1 \\ \vdots \\ x_m \end{pmatrix} = \begin{pmatrix} 1 \\ 0 \\ \vdots \\ 0 \end{pmatrix}.$$

Then $p(\varsigma)$ does not vanish on the unit circle. Moreover, if $d_m d_{m-1} > 0$ (respectively $d_m d_{m-1} < 0$) then $p(\varsigma)$ has P (respectively V) zeros inside the unit circle and V (respectively P) zeros outside the unit circle, counting multiplicities.

This theorem admits the following block generalization. Here and in the rest of the paper we deal with block matrices, where the blocks are of a fixed order r. We will also use the notation $I = (\delta_{ij})_{i,j=1}^{r}$.

Let $R = (R_{i-j})_{ij=0}^{m}$ be an invertible self-adjoint block Toeplitz matrix such that

$$(R^{-1})_{m,m} > 0 \ .$$

Let $a_0, \ldots, a_m$ be $m+1$ matrices of order r which solve the system of matrix equations

$$R_j a_0 + \ldots + R_{j-m} a_m = \delta_{j,0} I \qquad (j = 0, 1, \ldots, m) \ .$$

Then the determinant of the matrix polynomial

$$p(\varsigma) = a_m \varsigma^m + \cdots + a_1 \varsigma + a_0$$

has no zeros on the unit circle. Moreover, the sum of algebraic multiplicities of the zeros of det $(p(\varsigma))$ *inside the unit circle is equal to the number of negative eigenvalues of the principal minor $(R_{i-j})_{ij=0}^{m-1}$, counting multiplicities.*

This generalization is proved independently by D. Alpay and I. Gohberg (see [1]), and by I. Gohberg and L. Lerer (see [4]). Our methods here are similar in spirit to those of [4], while the proof in [1] is based on totally different ideas. In order to explain our aim it will be convenient to reformulate the previous statement in the following equivalent way.

Let $(R_{i-j})_{ij=-\infty}^{\infty}$ be a double infinite self-adjoint block Toeplitz matrix such that $R_{i-j} = 0$ if $|i-j| > m$, the matrix $R = (R_{i-j})_{ij=0}^{m}$ is invertible and $(R^{-1})_{m,m} > 0$. Let $(a_0, \ldots, a_m)^T$ be a block solution of the equation

$$\begin{pmatrix} R_0 & R_{-1} & \cdots & R_{-m} \\ R_1 & R_0 & \cdots & R_{-(m-1)} \\ \cdot & \cdot & \cdots & \cdot \\ R_m & R_{m-1} & \cdots & R_0 \end{pmatrix} \begin{pmatrix} a_0 \\ a_1 \\ \vdots \\ a_m \end{pmatrix} = \begin{pmatrix} I \\ 0 \\ \vdots \\ 0 \end{pmatrix} .$$

Then the lower triangular, double infinite, block Toeplitz matrix

$$\begin{pmatrix} \cdot & \cdot & \cdot & \cdot \\ \cdot & a_0 & 0 & \cdot \\ \cdot & a_1 & a_0 & \cdot \\ \cdot & \cdot & \cdot & \cdot \\ \cdot & a_m & a_{m-1} & \cdot \\ \cdot & 0 & a_m & \cdot \\ \cdot & \cdot & \cdot & \cdot \end{pmatrix}$$

is invertible. Moreover, the lower triangular, one side infinite, block Toeplitz matrix

$$\begin{pmatrix} a_0 & 0 & \cdot & \cdot & \cdot \\ a_1 & a_0 & \cdot & \cdot & \cdot \\ \cdot & \cdot & \cdot & \cdot & \cdot \\ a_m & a_{m-1} & \cdot & \cdot & \cdot \\ 0 & a_m & \cdot & \cdot & \cdot \\ \cdot & \cdot & \cdot & \cdot & \cdot \end{pmatrix}$$

is Fredholm, and its index is equal to the negative of the number of negative eigenvalues of the matrix $(R_{i-j})_{ij=0}^{m-1}$*, counting multiplicities.*

The main aim of this paper is to generalize this theorem for the nonstationary case, when the above matrices do not have the Toeplitz structure anymore. The following theorem is the main result of this paper. It may be viewed as a nonstationary generalization of the previous one.

THEOREM 1.1. *Let* $R = (R_{ij})_{ij=-\infty}^{\infty}$ *be a self-adjoint block matrix the entries of which,* R_{ij}*, are* $r \times r$ *complex matrices with the following properties:*

a) $R_{ij} = 0$ *if* $|i-j| > m$*, where* m *is a positive integer, and* $\sup_{ij} \|R_{ij}\| < +\infty$.

b) *The matrices* $(R_{ij})_{ij=n}^{n+m}$ *and* $(R_{ij})_{ij=n}^{n+m-1}$ $(n = 0, \pm 1, \ldots,)$ *are invertible and*

$$\sup_n (\|[(R_{ij})_{ij=n}^{n+m}]^{-1}\| \; , \quad \|[(R_{ij})_{ij=n}^{n+m-1}]^{-1}\|) < +\infty \; .$$

c) *The number of negative eigenvalues of the matrices* $(R_{ij})_{ij=n}^{n+m-1}$ $(n = 0, \pm 1, \ldots,)$ *does not depend on* n*, and*

$$((R_{ij})_{ij=n}^{n+m})_{m,m} > 0 \quad (n = 0, \pm 1, \ldots,) \; .$$

For every integer n*, let* $(a_{n+k,n})_{k=0}^{m}$ *be the solution of the system*

$$\sum_{k=0}^{m} R_{n+h,n+k} a_{n+k,n} = \delta_{0,h} I \;\; (h = 0, \ldots, m) \; , \tag{1.1}$$

and let $a_{ij} = 0$ *if* $i < j$ *or* $i > j+m$*. Then the matrix* $A = (a_{ij})_{ij=-\infty}^{\infty}$ *defines an invertible operator in* $\ell_r^2(\mathbb{Z})$*, the matrix* $G = (a_{ij})_{ij=0}^{\infty}$ *defines a Fredholm operator in* ℓ_r^2*, and the index of* G *is equal to the negative of the number of negative eigenvalues of* $(R_{ij})_{ij=n}^{n+m-1}$ *for any* $n = 0, \pm 1, \ldots,$ *counting multiplicities.*

Let us remark that in this theorem Ker $G = \{0\}$.

This paper is divided into four sections. The first is the introduction. The next section contains results from [2] and [3] which will be used in the proof of Theorem 1.1.

Section 3 is devoted to the construction of a nonstationary Stein equation from the data in Theorem 1.1. The proof of Theorem 1.1 appears in the last section.

2. PRELIMINARIES

The notion of dichotomy is connected to nonstationary Stein equations and band matrices. We begin by defining the dichotomy of a sequence of matrices.

Let $(A_n)_{n=-\infty}^{\infty}$ be a sequence of $h \times h$ invertible matrices. We view the matrices $(A_n)_{n=-\infty}^{\infty}$ as operating on row vectors on the left. A sequence of projections $(P_n)_{n=-\infty}^{\infty}$ in C^h, satisfying $\sup_n ||P_n|| < +\infty$, is called a left dichotomy for $(A_n)_{n=-\infty}^{\infty}$ if

$$P_n A_n = A_n P_{n+1} \qquad (n = 0, \pm 1, \cdots) ,$$

and if there are two positive numbers a and M, with $a < 1$, such that

$$||P_n A_n \cdots A_{n+j-1}|| \leq M a^j$$

and

$$||(I_h - P_n) A_{n-1}^{-1} \cdots A_{n-j}^{-1}|| < M a^j$$

where $n = 0, \pm 1, \cdots$; $j = 0, 1, 2, , \ldots$, and $I_h = (\delta_{ij})_{i,j=1}^{h}$. Such a pair (a, M) is called a bound of the dichotomy, and the constant number rank (P_n) is called the rank of the dichotomy. The paper [2] contains a description of dichotomy in this, and more general cases.

We will also use the following remark. Let $(A_n)_{n=-\infty}^{\infty}$ be a sequence of invertible matrices which admits a left dichotomy $(P_n)_{n=-\infty}^{\infty}$ with bound (a, M), and let $(K_n)_{n=-\infty}^{\infty}$ be a sequence of invertible matrices such that $||K_n|| \leq N$ and $||K_n^{-1}|| \leq N$ for every $n = 0, \pm 1, \cdots$. Then the sequence $(K_n^{-1} A_n K_{n+1})_{n=-\infty}^{\infty}$ admits the left dichotomy $(K_n^{-1} P_n K_n)_{n=-\infty}^{\infty}$ with bound (a, MN^2). This is a simple consequence of the definitions.

We now proceed to the definition of nonstationary Stein equations. A nonstationary Stein equation is an infinite set of matrix equations of the form

$$X_n - A_n^* X_{n+1} A_n = D_n \qquad (n = 0, \pm 1, \ldots) ,$$

where the given sequences $(A_n)_{n=-\infty}^{\infty}$ and $(D_n)_{n=-\infty}^{\infty}$, as well as the solution sequence $(X_n)_{n=-\infty}^{\infty}$, consist of $h \times h$ matrices. We say that the nonstationary Stein equation is backward positive if

$$D_n \geq 0 \qquad (n = 0, \pm 1, \ldots) ,$$

and if there exists a number $\varepsilon > 0$ and an integer $\ell > 0$ such that

$$D_n + A_n^* D_{n+1} A_n + \cdots + A_n^* \cdots A_{n+\ell-2}^* D_{n+\ell-1} A_{n+\ell-2} \cdots A_n \geq$$
$$\geq \varepsilon A_n^* \cdots A_{n+\ell-1}^* A_{n+\ell-1} \cdots A_n ,$$

for $n = 0, \pm 1, \ldots$. Such a pair (ε, ℓ) will be called a positivity bound. We refer to [2] for a more complete description of this and other types of positivity, and their applications.

If X is a finite self-adjoint matrix, we call the inertia of X the triple of integers (ν_+, ν_0, ν_-), where $\nu_0 = \dim \ker X$, and ν_+ (respectively ν_-) is the number of positive (respectively negative) eigenvalues of X. We say that a sequence of matrices $(X_n)_{n=-\infty}^{\infty}$ is of constant inertia (ν_+, ν_0, ν_-) if, for every $n = 0, \pm 1, \ldots$, the inertia of X_n is (ν_+, ν_0, ν_-).

Some connections between Stein equations and dichotomy are given by the following theorem, which is an immediate consequence of results of [2] (see Theorem 5.3 (condition g)), Lemma 7.1, and remark in Section 2, of [2]).

THEOREM 2.1. *Let*

$$X_n - A_n X_{n+1} A_n^* = D_n \qquad (n = 0, \pm 1, \ldots)$$

be a backward positive Stein equation, with positivity bound (ε, ℓ), *such that all* $(A_n)_{n=-\infty}^{\infty}$ *are invertible and* $\sup_n \|A_n^{-1}\| < +\infty$. *Assume that this equation has a bounded self-adjoint solution* $(X_n)_{n=-\infty}^{\infty}$ *of constant inertia* (ν_+, ν_0, ν_-). *Then* $\nu_0 = 0$ *and the sequence* $(A_n)_{n=-\infty}^{\infty}$ *admits a left dichotomy of rank* ν_+, *with a bound given by*

$$(a, M) = \left((1 + \frac{1}{\alpha})^{-\frac{1}{2\ell}},\ (2\alpha + 2)^2 \right) ,$$

where $\alpha = 2 \sup_n \|X_n\| (1 + \sup_n \|A_n^{-1}\|^{2\ell}) / \varepsilon$.

A block matrix $(t_{ij})_{ij=-\infty}^{\infty}$ is called m-band if $t_{ij} = 0$ whenever $|i - j| > m$. Let $A = (a_{ij})_{ij=-\infty}^{\infty}$ be a lower triangular m-band block matrix. If $a_{n,n}$ and $a_{n+m,n}$ are invertible for $n = 0, \pm 1, \ldots$, then A is called regular. In this case, we associate with A its companion sequence $(C_n)_{n=-\infty}^{\infty}$, which is defined as follows

$$C_n = \begin{pmatrix} 0 & 0 & \cdot & \cdot & \cdot & 0 & -a_{n,n} a_{n+m,n}^{-1} \\ I & 0 & \cdot & \cdot & \cdot & 0 & -a_{n+1,n} a_{n+m,n}^{-1} \\ \cdot & \cdot & \cdot & \cdot & \cdot & \cdot & \cdot \\ 0 & 0 & \cdot & \cdot & \cdot & I & -a_{n+m-1,n} a_{n+m,n}^{-1} \end{pmatrix} \qquad (n = 0, \pm 1, \cdots) . \tag{2.1}$$

We will use Theorem 1.1 of [3] which, for convenience, is restated here as Theorem 2.2. An infinite block matrix $(a_{ij})_{ij=-\infty}^{\infty}$ is called bounded if it defines a bounded operator in $\ell_r^2(\mathbb{Z})$.

THEOREM 2.2. *Let $A = (a_{ij})_{ij=-\infty}^{\infty}$ be a lower triangular regular m-band bounded block matrix such that $\sup_n \|a_{n,n}^{-1}\| < +\infty$. The matrix A represents an invertible operator in $\ell_r^2(\mathbb{Z})$ if and only if its companion sequence admits a left dichotomy. If A is invertible in $\ell_r^2(\mathbb{Z})$ then $\|A^{-1}\| \leq 2M^2(1-a)^{-1}(\sup_n \|a_{n,n}^{-1}\|)$, where (a, M) is a bound of the dichotomy, the matrix $G = (a_{ij})_{ij=0}^{\infty}$ represents a Fredholm operator in ℓ_r^2, and the index of G is equal to the negative of the rank of the dichotomy.*

3. BAND MATRICES AND NONSTATIONARY STEIN EQUATIONS.

Let $R = (R_{ij})_{ij=\infty}^{\infty}$ be a self-adjoint block m-band matrix. For each integer n we define the following matrices:

$$H_n = \begin{pmatrix} R_{n+1,n} & \cdot & \cdot & \cdot & R_{n+1,n+m-1} \\ \cdot & \cdot & \cdot & \cdot & \cdot \\ \cdot & \cdot & \cdot & \cdot & \cdot \\ R_{n+m,n} & \cdot & \cdot & \cdot & R_{n+m,n+m-1} \end{pmatrix}, \tag{3.1}$$

$$K_n = \begin{pmatrix} R_{n,n} & \cdot & \cdot & \cdot & R_{n,n+m-1} \\ \cdot & \cdot & \cdot & \cdot & \cdot \\ \cdot & \cdot & \cdot & \cdot & \cdot \\ R_{n+m-1,n} & \cdot & \cdot & \cdot & R_{n+m-1,n+m-1} \end{pmatrix}, \tag{3.2}$$

and

$$L_n = \begin{pmatrix} R_{n,n} & \cdot & \cdot & \cdot & R_{n,n+m} \\ \cdot & \cdot & \cdot & \cdot & \cdot \\ \cdot & \cdot & \cdot & \cdot & \cdot \\ \cdot & \cdot & \cdot & \cdot & \cdot \\ R_{n+m,n} & \cdot & \cdot & \cdot & R_{n+m,n+m} \end{pmatrix}. \tag{3.3}$$

The following is the main result of this section.

THEOREM 3.1. *Let $R = (R_{ij})_{ij=-\infty}^{\infty}$ be an m-band self-adjoint block matrix. Assume that the matrices H_n, K_n and L_n are invertible for every $n = 0, \pm 1, \ldots$. Then the nonstationary Stein equation*

$$X_n - (K_n H_n^{-1}) X_{n+1} (K_n H_n^{-1})^* = (K_n H_n^{-1}) E_n (K_n H_n^{-1})^* \qquad (n = 0, \pm 1, \ldots) \tag{3.4}$$

where

$$E_n = \begin{pmatrix} 0 & \cdot & \cdot & \cdot & 0 & 0 \\ \cdot & \cdot & \cdot & \cdot & \cdot & \cdot \\ 0 & \cdot & \cdot & \cdot & 0 & 0 \\ 0 & \cdot & \cdot & \cdot & 0 & ((L_n^{-1})_{m,m})^{-1} \end{pmatrix} \qquad (n = 0, \pm 1, \ldots) \tag{3.5}$$

admits the solution

$$X_n = -K_n \qquad (n = 0, \pm 1, \ldots) .$$

Moreover, if

(3.6) $$(L_n^{-1})_{m,m} > 0 \qquad (n = 0, \pm 1, \ldots) ,$$

and

(3.7) $$N = \sup_n \{||K_n^{-1}||, \ ||L_n^{-1}||, \ ||L_n||\} < \infty ,$$

then the nonstationary Stein equation (3.4) *is backward positive, with positivity bound* (ε, ℓ) *given by*

(3.8) $$\varepsilon = \left(\frac{1}{128m^2 N^{(4m+9)}}\right)^{2^{m+1}} ,$$

and

$$\ell = m .$$

In the proof we will need the following lemma.

LEMMA 3.2. *Let be given* $2m-1$ *block matrices* $U_1, \ldots, U_{m-1}, V_1, \ldots, V_m$*, of block order* $m \times m$*, which have the following form*

(3.9) $$U_j = \begin{pmatrix} 0 & I & \cdot & \cdot & \cdot & 0 & 0 \\ 0 & 0 & \cdot & \cdot & \cdot & 0 & 0 \\ \cdot & \cdot & \cdot & \cdot & \cdot & \cdot & \cdot \\ 0 & 0 & \cdot & \cdot & \cdot & 0 & I \\ * & * & \cdot & \cdot & \cdot & * & * \end{pmatrix} \qquad (j = 1, \ldots, m-1) ,$$

where $*$ *denotes an* $r \times r$ *matrix, and*

(3.10) $$V_j = \begin{pmatrix} 0 & \cdot & \cdot & \cdot & 0 & 0 \\ 0 & \cdot & \cdot & \cdot & 0 & 0 \\ \cdot & \cdot & \cdot & \cdot & \cdot & \cdot \\ 0 & \cdot & \cdot & \cdot & 0 & 0 \\ 0 & \cdot & \cdot & \cdot & 0 & \theta_j \end{pmatrix} \qquad (j = 1, \ldots, m) ,$$

and such that for some $L \geq 1$ *and* $L_1 > 0$

(3.11) $$||U_j|| \leq L \ ; \quad \frac{1}{L_1} I \leq \theta_j \leq L_1 I \quad (j = 1, \ldots, m) .$$

Then

$$V_1 + U_1V_2U_1^* + \cdots + U_1 \cdots U_{m-1}V_mU_{m-1}^* \cdots U_1^* \geq \varepsilon I \tag{3.12}$$

where

$$\varepsilon = \left(\frac{1}{16m^2L_1^3L^{2m}}\right)^{2^{m+1}} . \tag{3.13}$$

PROOF. We will prove the lemma by contradiction. Let $x = (x_1, \ldots, x_m)$ be a block row vector, such that $||x|| = 1$, and assume that

$$x(V_1 + U_1V_2U_1^* + \cdots + U_1 \cdots U_{m-1}V_mU_{m-1}^* \cdots U_1^*)x^* < \varepsilon .$$

This implies, in particular, that

$$xU_1 \cdots U_jV_{j+1}U_j^* \cdots U_1^*x^* \leq \varepsilon \qquad (j = 0, \ldots, m-1) . \tag{3.14}$$

We will denote by S the $m \times m$ block upper shift matrix

$$S = \begin{pmatrix} 0 & I & \cdot & \cdot & \cdot & 0 & 0 \\ 0 & 0 & \cdot & \cdot & \cdot & 0 & 0 \\ \cdot & \cdot & \cdot & \cdot & \cdot & \cdot & \cdot \\ 0 & 0 & \cdot & \cdot & \cdot & 0 & I \\ 0 & 0 & \cdot & \cdot & \cdot & 0 & 0 \end{pmatrix} .$$

The previous inequalities imply that

$$\begin{aligned} &xS^jV_{j+1}S^{j^*}x^* + x(U_1 \cdots U_j - S^j)V_{j+1}(xS^j)^* + \\ &\quad + xU_1 \cdots U_jV_{j+1}(U_1 \cdots U_j - S^j)^*x^* \leq \varepsilon \qquad (j = 1, \ldots, m-1) , \end{aligned} \tag{3.15}$$

On the other hand, the identities

$$U_1 \cdots U_j - S^j = (U_1 - S)U_2 \cdots U_j + S(U_2 - S)U_3 \cdots U_j + \cdots + S^{j-1}(U_j - S)$$

where $j = 1, \ldots, m-1$, and the inequalities

$$||xS^k(U_j - S)|| \leq L||x_{m-k}|| \qquad (j = 1, \ldots, m-1 ; \quad k = 0, \ldots, m-1) ,$$

lead to

$$||x(U_1 \cdots U_j - S^j)|| \leq L^j||x_m|| + \cdots + L||x_{m-j+1}|| \qquad (j = 1, \ldots, m-1) .$$

Since $L \geq 1$, then we obtain

$$||x(U_1 \cdots U_j - S^j)|| \leq L^m(||x_m|| + \cdots + ||x_{m-j+1}||) \quad (j = 1, \ldots, m-1) .$$

Using these inequalities and $||x|| = 1$, it is easily seen that the inequalities (3.15) lead to

$$xS^jV_{j+1}S^{j*}x^* \leq \varepsilon + 2L_1L^{2m}(||x_m|| + \cdots + x_{m-j+1}||) \qquad (j = 1, \ldots, m-1) .$$

However, by (3.11) we have

$$xS^jV_{j+1}S^{j*}x^* = x_{m-j}\theta_{j+1}x^*_{m-j} \geq \frac{1}{L_1}||x_{m-j}||^2 \qquad (j = 1, \ldots, m-1) .$$

Combining the two previous sets of inequalities we obtain

$$(3.16) \quad ||x_{m-j}|| \leq (\varepsilon L_1 + 2L_1^2L^{2m}(||x_m|| + \cdots + ||x_{m-j+1}||))^{1/2} \qquad (j = 1, \ldots, m-1) .$$

Moreover, (3.14),with $j = 0$, and (3.11) show that $\frac{1}{L_1}||x_m||^2 \leq x_m\theta_1x^*_m \leq \varepsilon$,and therefore

$$(3.17) \qquad ||x_m|| \leq (\varepsilon L_1)^{1/2} .$$

Using these inequalities, we will obtain a contradiction.

Denote $\alpha = (\varepsilon L_1)^{1/2}$, $\beta = \sqrt{2}L_1L^m\sqrt{m}$, and define $m-1$ numbers $\lambda_0, \ldots, \lambda_{m-1}$ via the following recursion, $\lambda_0 = \alpha$ and

$$\lambda_k = \alpha + \beta \cdot (\lambda_{k-1})^{1/2} \qquad (k = 1, \ldots, m-1) .$$

Since $\lambda_{k-1} \geq \alpha$ and $\alpha < 1 < \beta$, we have $\beta(\lambda_{k-1})^{1/2} \geq \alpha^{1/2} \geq \alpha$, and therefore,

$$\lambda_k \leq 2\beta(\lambda_{k-1})^{1/2} \qquad (k = 1, \ldots, m-1) .$$

Thus, by a simple iteration, we obtain

$$\lambda_k \leq 2\beta(\lambda_{k-1})^{1/2} \leq (2\beta)^{1+\frac{1}{2}}(\lambda_{k-2})^{1/4} \leq \cdots \leq (2\beta)^2\lambda_0^{1/2^k} \qquad (k = 1, \ldots, m-1) .$$

Taking into account that $\lambda_0 = \alpha = (\varepsilon L_1)^{1/2}$ and $L_1 \geq 1$, this leads to

$$(3.18) \qquad \lambda_k \leq 8L_1^3L^{2m} \cdot m\, \varepsilon^{\frac{1}{2(m+1)}} \qquad (k = 0, \ldots, m-1) .$$

In particular, it follows that $\lambda_k \leq 1/2 \quad (k = 0, \ldots, m-1)$, and thus $(\lambda_k)^{1/2} \geq \lambda_k$. Since $\beta \geq 1$, we obtain from the recursion above

$$\lambda_k \geq \lambda_{k-1} \qquad (k = 1, \ldots, m-1) .$$

Let us now prove by induction that

$$\lambda_k \geq \max(||x_m||, \ldots, ||x_{m-k}||) \qquad (k = 0, \ldots, m-1) \ . \tag{3.19}$$

This follows from (3.17) for $k = 0$. Assuming that (3.19) holds for $k-1$ we obtain from (3.16)

$$||x_{m-k}|| \leq (\alpha^2 + \beta^2 \lambda_{k-1})^{1/2} \leq \alpha + \beta(\lambda_{k-1})^{1/2} = \lambda_k \ .$$

This inequality and $\lambda_k \geq \lambda_{k-1} \geq \max(||x_m||, \ldots, ||x_{m-k+1}||)$ imply (3.19).

Now, inequality (3.18) with $k = m-1$, combined with (3.19), shows that

$$||x|| \leq \sum_{i=1}^{m} ||x_i|| \leq 8L^3 L_1^{2m} m^2 \varepsilon^{\frac{1}{2(m+1)}} = \frac{1}{2} \ ,$$

which contradicts the assumption $||x|| = 1$. □

PROOF OF THEOREM 3.1. Let n be an integer. Note that the i'th row of H_n equals the $(i+1)$'th row of K_n, for $i = 1, \ldots, m-1$. Therefore, the matrix $H_n K_n^{-1}$ has the following form

$$H_n K_n^{-1} = \begin{pmatrix} 0 & I & \cdot & \cdot & \cdot & 0 & 0 \\ 0 & 0 & \cdot & \cdot & \cdot & 0 & 0 \\ \cdot & \cdot & \cdot & \cdot & \cdot & \cdot & \cdot \\ 0 & 0 & \cdot & \cdot & \cdot & 0 & I \\ * & * & \cdot & \cdot & \cdot & * & * \end{pmatrix} . \tag{3.20}$$

Thus, $H_n K_n^{-1} H_n^*$ has the following form

$$H_n K_n^{-1} H_n^* = \begin{pmatrix} R_{n+1,n+1} & \cdot & \cdot & \cdot & R_{n+1,n+m} \\ \cdot & \cdot & \cdot & \cdot & \cdot \\ R_{n+m-1,n+1} & \cdot & \cdot & \cdot & R_{n+m-1,n+m} \\ * & \cdot & \cdot & \cdot & * \end{pmatrix} .$$

Since both K_{n+1} and $H_n K_n^{-1} H_n^*$ are self-adjoint, we conclude that there exist matrices μ_n $(n = 0, \pm 1, \ldots)$ such that

$$K_{n+1} - H_n K_n^{-1} H_n^* = \begin{pmatrix} 0 & \cdot & \cdot & \cdot & 0 & 0 \\ \cdot & \cdot & \cdot & \cdot & \cdot & \cdot \\ 0 & \cdot & \cdot & \cdot & 0 & 0 \\ 0 & \cdot & \cdot & \cdot & 0 & \mu_n \end{pmatrix} . \tag{3.21}$$

By inspection, it follows that

$$\mu_n = R_{n+m,n+m} - (R_{n+m,n} \cdots R_{n+m,n+m-1}) K_n^{-1} \begin{pmatrix} R_{n,n+m} \\ \cdots \\ R_{n+m-1,n+m} \end{pmatrix} . \tag{3.22}$$

This expression shows that μ_n is the Schur complement of K_n in L_n. It is well known, and easy to verify, that

$$(3.23) \qquad \mu_n^{-1} = (L_n^{-1})_{m,m} \qquad (n = 0, \pm 1, \ldots) .$$

The first part of the theorem is a consequence of equalities (3.21) and (3.23).

In order to prove that equation (2.4) is backward positive we must show that

$$\begin{aligned} &T_n E_n T_n^* + T_n(T_{n+1}E_{n+1}T_{n+1}^*)T_n^* + \cdots + \\ &\quad + T_n \cdots T_{n+m-2}(T_{n+m-1}E_{n+m-1}T_{n+m-1}^*)T_{n+m-2}^* \cdots T_n^* \geq \\ &\quad \geq \varepsilon T_n \cdots T_{n+m-1}T_{n+m-1}^* \cdots T_n^* \qquad (n = 0, \pm 1, \ldots) , \end{aligned}$$

where $T_n = K_n H_n^{-1}$ $(n = 0, \pm 1, \ldots,)$. This inequality is equivalent to

$$(3.24) \qquad \begin{aligned} &E_{n+m-1} + T_{n+m-1}^{-1} E_{n+m-2} T_{n+m-1}^{-1*} + \\ &\cdots + T_{n+m-1}^{-1} \cdots T_{n+1}^{-1} E_n T_{n+1}^{-1*} \cdots T_{n+m-1}^{-1*} \geq \varepsilon\, I \qquad (n = 0, \pm 1, \ldots) . \end{aligned}$$

However, by (3.7), (3.22) and (3.23) we have

$$||\mu_n|| \leq N + N^3 \quad , \quad ||\mu_n^{-1}|| \leq N ,$$

Since $\mu_n > 0$, these inequalities imply that

$$(3.25) \qquad \frac{1}{2N^3} I \leq \mu_n \leq 2N^3 I \qquad (n = 0, \pm 1, \ldots) ,$$

where we have used $N \geq 1$. Moreover, it is clear by (3.7) that

$$(3.26) \qquad ||T_n^{-1}|| = ||H_n K_n^{-1}|| \leq N^2 \qquad (n = 0, \pm 1, \ldots) .$$

Define $U_j = T_{n+m-j}^{-1}$ $(j = 1, \ldots, m-1)$, $V_j = E_{n+m-j}$ $(j = 1, \ldots, m)$, $L = N^2$, and $L_1 = 2N^3$. It follows from (3.5) and (3.20) that U_j and V_j have the structures as in (3.9) and (3.10). Moreover, inequalities (3.11) follow from (3.25) and (3.26). Thus we can apply Lemma 3.1. Inequality (3.24), with ε given by (3.8), follows from (3.12) and (3.13). □

4. PROOF OF THE MAIN RESULT.

PROOF OF THEOREM 1.1. We keep the notation given by (3.1) – (3.3), and set

$$N = \sup_n (||K_n^{-1}||, \; ||[L_n^{-1}||, \; ||L_n||) .$$

We first prove the theorem with the additional condition that all the matrices H_n $(n = 0, \pm1, \ldots,)$ are invertible. Theorems 3.1 and 2.1 imply that the sequence $(K_n H_n^{-1})_{n=-\infty}^{\infty}$ admits a left dichotomy of rank ν_- with a bound given by

$$(a, M) = \left((1 + \frac{1}{\alpha})^{\frac{-1}{2m}},\ (2\alpha + 2)^2\right) , \tag{4.1}$$

where

$$\alpha = 2N(1 + N^{4m})(128m^2 N^{4m+9})^{2^{m+1}} . \tag{4.2}$$

Let $(C_n)_{n=-\infty}^{\infty}$ be the companion sequence of A. It follows easily by (2.1) that $H_n C_n = K_{n+1}$ $(n = 0, \pm1, \ldots,)$. Therefore $C_n = H_n^{-1} K_{n+1} = K_n^{-1}(K_n H_n^{-1}) K_{n+1}$ $(n = 0, \pm1, \ldots,)$. Hence, $(C_n)_{n=-\infty}^{\infty}$ admits a left dichotomy of rank ν_- with bound (a, MN^2) (we use here the previous paragraph and a remark from Section 2).

Now note that $a_{n,n}$ is the Schur complement of K_{n+1} in L_n, namely

$$a_{n,n}^{-1} = R_{n,n} - (R_{n,n+1}, \ldots, R_{n,n+m}) K_{n+1}^{-1} \begin{pmatrix} R_{n+1,n} \\ \cdots \\ R_{n+m,n} \end{pmatrix} .$$

Therefore,

$$||a_{n,n}^{-1}|| \le N + N^3 \qquad (n = 0, \pm1, \ldots,) . \tag{4.3}$$

Theorem 1.1, in this case, follows immediately from Theorem 2.2. In addition, Theorem 2.2 shows that

$$||A^{-1}|| \le 2(MN^2)^2 (1 - a)^{-1} (N + N^3) , \tag{4.4}$$

where a and M are given by (4.1) and (4.2).

We now return to the general case, where the matrices H_n $(n = 0, \pm1, \ldots,)$ are no longer assumed to be invertible.

Let $(\varepsilon_k)_{k=-\infty}^{\infty}$ be a sequence of real numbers such that $\lim_{k \to +\infty} \varepsilon_k = 0$, $|\varepsilon_k| < \frac{1}{4N}$ $(k = 0, \pm1, \cdots)$, and the matrices $(R_{ij} + \varepsilon_k \delta_{i+1,j} I + \varepsilon_k \delta_{i,j+1} I)_{i=n+1,j=n}^{n+m,n+m-1}$ are invertible for every $n, k = 0, \pm1, \ldots$. Such a sequence exists because, for every n, the determinant

$$\det[(R_{ij} + \lambda\delta_{i+1,j} I + \lambda\delta_{i,j+1} I)_{i=n+1,j=n}^{n+m,n+m-1}] ,$$

is, as a function of λ, a monic polynomial of degree rm , and has therefore rm roots.

We will denote

$$R_{ij;k} = R_{ij} + \varepsilon_k \delta_{i+1,j} I + \varepsilon_k \delta_{i,j+1} I \qquad (i,j,k = 0, \pm 1, \cdots) ,$$

and

$$H_{n;k} = (R_{ij;k})_{i=n+1,j=n}^{n+m,n+m-1} , \quad K_{n;k} = (R_{ij;k})_{i,j=n}^{n+m-1} , \quad L_{n;k} = (R_{ij;k})_{i,j=n}^{n+m} ,$$

for $k, n = 0, \pm 1, \ldots$. All these matrices are invertible and the following inequalities hold

$$||K_{n;k}^{\pm 1}|| \leq 2N ; \quad ||L_{n;k}^{\pm 1}|| \leq 2N \qquad (k, n = 0, \pm 1, \cdots) .$$

In addition all the matrices K_n $(n = 0, \pm 1, \ldots)$ and $K_{n;k}(k, n = 0, \pm 1, \cdots)$ have the same inertia $(\nu_+, 0, \nu_-)$, and all the matrices $H_{n,k}$ $(k, n = 0, \pm 1, \ldots,)$ are invertible.

For all integers $k, n = 0, \pm 1, \ldots,$ let $(a_{n+t,n;k})_{t=0}^{m}$ be the first block column of the inverse of $L_{n,k}$. Thus, the matrices $(a_{n+t,n;k})_{t=0}^{m}$ are the solutions of the following system

$$R_{n+j,n;k} a_{n,n;k} + \cdots + R_{n+j,n+m;k} a_{n+m,n;k} = \delta_{j,0} I \qquad (j = 0, \ldots, m) .$$

We also set $a_{ij;k} = 0$ for $i < j$ or $i > j + m$, and define

$$A_k = (a_{ij;k})_{ij=-\infty}^{\infty} \quad \text{and} \quad G_k = (a_{ij;k})_{ij=0}^{\infty} \qquad (k = 0, \pm 1, \cdots) .$$

Then

$$\lim_{k \to \infty} A_k = A \quad \text{and} \quad \lim_{k \to \infty} G_k = G , \tag{4.5}$$

in the norm of operators.

We can apply the first part of the proof to A_k $(k = 1, 2, \ldots,)$. Therefore, for every $k = 1, 2, \ldots,$ A_k is invertible in $\ell_r^2(\mathbb{Z})$, G_k is Fredholm with

$$\text{index } G_k = -\nu_- \qquad (k = 1, 2, \ldots,) , \tag{4.6}$$

and we have a uniform estimate

$$||A_k^{-1}|| \leq B \qquad (k = 1, 2, \ldots,) . \tag{4.7}$$

Here B is a constant independent of k, which is obtained by substituting $2N$ with N in (4.1), (4.2) and (4.4).

By (4.7), $\sup_k ||A_k^{-1}|| < +\infty$. Therefore the first limit in (4.5) implies that A is invertible. Consequently G is Fredholm. Finally, the second limit in (4.5), and equality (4.6) imply that index $G = -\nu_-$. □

REFERENCES.

[1] D. Alpay and I. Gohberg, On Orthogonal Matrix Polynomials, this volume.

[2] A. Ben-Artzi and I. Gohberg, Inertia Theorems for Nonstationary Discrete Systems and Dichotomy, to appear in Linear Algebra and its Applications.

[3] A. Ben-Artzi and I. Gohberg, Fredholm Properties of Band Matrices and Dichotomy. Operator Theory: Advances and Applications, Vol. 32, Topics in Operator Theory. Constantin Apostol Memorial Issue, Birkhauser Verlag, 1988.

[4] I. Gohberg and L. Lerer, Matrix Generalizations of M.G. Krein Theorems on Orthogonal Polynomials, this volume.

Raymond and Beverly Sackler Faculty of Exact Sciences
School of Mathematical Sciences
Tel-Aviv University, Israel

Operator Theory:
Advances and Applications, Vol. 34

HERMITIAN BLOCK TOEPLITZ MATRICES, ORTHOGONAL POLYNOMIALS, REPRODUCING KERNEL PONTRYAGIN SPACES, INTERPOLATION AND EXTENSION

Harry Dym*

A largely expository account of the theory of matrix orthogonal polynomials associated with Hermitian block Toeplitz matrices, including assorted recursions, algorithms and zero properties is prepared, using structured reproducing kernel Pontryagin spaces as a key tool. A number of closely related problems of interpolation and extension are also studied.

CONTENTS

---

* The author would like to acknowledge with thanks Renee and Jay Weiss for endowing the chair which supported this research.

1. INTRODUCTION

This paper is a largely expository account of the theory of $p \times p$ matrix polynomials associated with Hermitian block Toeplitz matrices and some related problems of interpolation and extension. Perhaps the main novelty is the use of reproducing kernel Pontryagin spaces to develop parts of the theory in what hopefully the reader will regard as a reasonably lucid way. The topics under discussion are presented in a series of short sections, the headings of which give a pretty good idea of the overall contents of the paper. The theory is a rich one and the present paper in spite of its length is far from complete. The author hopes to fill in some of the gaps in future publications.

The story begins with a given sequence $h_{-n}, \ldots, h_n$ of $p \times p$ matrices with $h_{-j} = h_j^*$ for $j = 0, \ldots, n$. We let

$$H_k = \begin{bmatrix} h_0 & \cdots & h_{-k} \\ \vdots & & \vdots \\ h_k & \cdots & h_0 \end{bmatrix}, \qquad k = 0, \ldots, n, \tag{1.1}$$

denote the Hermitian block Toeplitz matrix based on $h_0, \ldots, h_k$ and shall denote its inverse H_k^{-1} by

$$\Gamma_k = \left[\gamma_{ij}^{(k)}\right]_{i,j=0}^{k}, \qquad k = 0, \ldots, n, \tag{1.2}$$

whenever H_k is invertible.

In the present study we shall always take H_n invertible and shall make extensive use of the polynomials

$$A_n(\lambda) = \sum_{i=0}^{n} \lambda^i \gamma_{i0}^{(n)} \quad \text{and} \quad C_n(\lambda) = \sum_{i=0}^{n} \lambda^i \gamma_{in}^{(n)} \tag{1.3}$$

of the first kind and the polynomials

$$A_n^o(\lambda) = 2I_p - \sum_{i=0}^{n} p_i(\lambda)\gamma_{i0}^{(n)} \quad \text{and} \quad C_n^o(\lambda) = \sum_{i=0}^{n} p_i(\lambda)\gamma_{in}^{(n)} \tag{1.4}$$

of the second kind, wherein

$$p_i(\lambda) = \lambda^i h_0^* + 2\sum_{s=1}^{i} \lambda^{i-s} h_s^* . \tag{1.5}$$

In the special case that $H_0, \ldots, H_n$ are all invertible, then the polynomials of (1.3) and (1.4) can be defined in the same way for every integer j, $j = 0, \ldots, n$, and it turns out that the columns of $\begin{bmatrix} C_j \\ C_j^o \end{bmatrix}$ are orthogonal to the columns of $\begin{bmatrix} C_i \\ C_i^o \end{bmatrix}$ with respect to the J_1 inner product for $i \neq j$:

$$< J_1 \begin{bmatrix} C_j \\ C_j^o \end{bmatrix} \xi \, , \, \begin{bmatrix} C_i \\ C_i^o \end{bmatrix} \eta >= \begin{cases} 0 & \text{if } i \neq j \\ \eta^* 2\gamma_{jj}^{(j)} \xi & \text{if } i = j \, , \end{cases} \tag{1.6}$$

$i, j = 0, \ldots, n$, for every choice of ξ and η in $\mathbb{C}^p$. Now, if

$$\Phi(\lambda) = \sum_{j=0}^{\infty} a_j \lambda^j$$

is any $p \times p$ matrix valued analytic function on the open unit disc $\mathbb{D}$ with

$$a_0 = h_0 \text{ and } a_j = 2h_j \ , \quad j = 1, \ldots, n$$

and say (for simplicity's sake)

$$\sum_{j=0}^{\infty} |a_j| < \infty \ ,$$

then it is readily checked that

$$C_j^o = \underline{\underline{p}}\Phi^* C_j \ , \qquad j = 0, \ldots, n \ , \tag{1.7}$$

in which $\underline{\underline{p}}$ designates the orthogonal projection onto H_p^2 (as is explained below), and hence that (1.6) reduces to

$$\frac{1}{2} < (\Phi + \Phi^*) C_j \xi, C_i \eta >= \begin{cases} 0 & \text{if } i \neq j \\ \eta^* \gamma_{jj}^{(j)} \xi & \text{if } i = j \ , \end{cases} \tag{1.8}$$

$i, j = 0, \ldots, n$. Formula (1.8) exhibits the C_j as orthogonal polynomials with respect to the density $(\Phi + \Phi^*)/2$ based on the "interpolant" Φ. If, as happens in the case that H_n is positive definite, $\det \Phi(\lambda) \neq 0$ for all points λ in the closed unit disc, then the recipe (1.7) can be inverted to obtain

$$C_j = \underline{\underline{p}}\{\Phi^*\}^{-1} C_j^o \tag{1.9}$$

and then (1.6) leads to the auxiliary formula

$$\frac{1}{2} < (\Phi^{-1} + \Phi^{*-1}) C_j^o \xi, C_i^o \eta >= \begin{cases} 0 & \text{if } i \neq j \\ \eta^* \gamma_{jj}^{(j)} \xi & \text{if } i = j \ , \end{cases} \tag{1.10}$$

A more careful analysis of (1.8) reveals the fact that the columns of C_j belong to the span of $\{\lambda^i \xi : \ i = 0, \ldots, j, \ \xi \in \mathbb{C}^p\}$ and are orthogonal to the span of $\{\lambda^i \xi : \ i = 0, \ldots, j-1, \ \xi \in \mathbb{C}^p\}$ with respect to the indicated indefinite inner product based on the matrix density $(\Phi + \Phi^*)/2$ (which will always be taken invertible, but not necessarily positive definite) and hence the C_j are often referred to as the forward innovation polynomials. Much the same sort of analysis leads to the conclusion that the columns of the matrix polynomials $\lambda^{n-j} A_j(\lambda)$ belong to the span of $\{\lambda^i \xi : \ i = n - j, \ldots, n, \ \xi \in \mathbb{C}^p\}$ and are orthogonal to the span of $\{\lambda^i \xi : \ i = n-j+1, \ldots, n, \ \xi \in \mathbb{C}^p\}$

with respect to the matrix density $(\Phi + \Phi^*)/2$. Correspondingly the matrix polynomials $\lambda^{n-j}A_j(\lambda)$ are backwards innovations polynomials. The main formulas are

$$< J_1\lambda^{n-j}\begin{bmatrix} A_j \\ A_j^o \end{bmatrix}\xi\ ,\ \lambda^{n-i}\begin{bmatrix} A_i \\ A_i^o \end{bmatrix}\eta\ >= \begin{cases} 0 & \text{if } i \neq j \\ 2\eta^*\gamma_{00}^{(j)}\xi & \text{if } i = j\ , \end{cases} \tag{1.11}$$

$$\frac{1}{2} < (\Phi + \Phi^*)\lambda^{n-j}A_j\xi,\ \lambda^{n-i}A_i\eta >= \begin{cases} 0 & \text{if } i \neq j \\ \eta^*\gamma_{00}^{(j)}\xi & \text{if } i = j\ , \end{cases} \tag{1.12}$$

for $i, j = 0, \ldots, n$ and every choice of ξ and η in $\mathbb{C}^p$.

Other characterizations and properties of these polynomials, and some related spaces will appear in the sequel more or less as indicated in the table of contents.

The literature on the orthogonal polynomials associated with Toeplitz matrices is vast. Most of the early work focused on the case where H_n is positive definite with scalar entries. For an extensive bibliography and a nice survey including some continuous analogues (which originate with Krein [K1] and have since been widely extended in a fundamental paper by Krein and Langer [KL]) see Hirschman [H2]. The latter is nicely complemented by Kailath, Vieira and Morf [KVM]; for applications to engineering and another extensive bibliography see also [Ka]. Among the earlier references particular attention should be paid to the classical monographs of Szegö and Geronimus as well as to the papers of Baxter and Hirschman.

The first systematic studies of orthogonal matrix polynomials associated with positive definite H_n seem to have been carried out by Delsarte, Genin and Kampe [DGK1] and Youla-Kazanjian [YK] in 1978; see also Morf, Vieira and Kailath [MVK]. A large part of these papers deal with recursion formulas. However, it turns out that most of these go through in one form or another when $H_0, \ldots, H_n$ are invertible, block Toeplitz Hermitian matrices; Sections 13 and 14 give a pretty good sample of the kind of results which are available under these assumptions. An earlier study which focused on inversion formulas for non Hermitian block Toeplitz matrices was undertaken by Gohberg and Heinig [GH]. This paper was partially motivated by the work of Hirschman [H1]. For the development of a number of recursion formulas in this setting and a good bibliography, see Fuhrmann [F].

The present paper evolved from the draft of a chapter which was originally written to illustrate certain reproducing kernel Hilbert spaces methods in a planned expanded version of [D], under the assumption that the block Toeplitz matrix H_n based on the given data $h_{-n}, \ldots, h_n$ was positive definite. Perhaps the main achievement of the original effort was the representation formulas for the solution of the covariance extension problem described in Section 10 (the formulas themselves, in a somewhat different setting, were first obtained by different methods by Youla [Y]). Interest in generalizing the interpolation methods which lead to these formulas to the indefinite case was largely sparked by a paper of Delsarte, Genin and Kampe [DGK2], which treated a number of interpolation problems, for Hermitian Toeplitz matrices H_n with scalar entries. Matrix versions of many of their representation formulas, in terms of the polynomials introduced in (1.3) and (1.4), are given in Sections 7 and 8. Matrix versions of some of their other results will be considered elsewhere.

Section 11 presents a new proof of a generalization to the matrix case by Alpay and Gohberg [AG] of a theorem of Krein [K2] on the number of roots of the orthogonal polynomial $C_n(\lambda)$ inside $\mathbb{D}$.

Finally, a few words on notation. The symbol $\mathbb{C}$ denotes the complex numbers, whereas $\mathbb{C}^{j\times k}$ stands for the space of $j \times k$ matrices with complex entries and $\mathbb{C}^j$ is short for $\mathbb{C}^{j\times 1}$; $\mathbb{D} = \{\lambda \in \mathbb{C} : |\lambda| < 1\}$, $\mathbb{T} = \{\lambda \in \mathbb{C} : |\lambda| = 1\}$ and $\mathbb{E} = \{\lambda \in \mathbb{C} \cup \{\infty\} : 1 < |\lambda| \leq \infty\}$.

If A is a $j \times k$ matrix, then $|A|$ stands for its maximum s-number, and A^* its conjugate transpose. If $A = A^*$, then $\mu_+(A)$, $\mu_-(A)$ and $\mu_0(A)$ denote the number of positive, negative and zero eigenvalues of A, respectively. The triple (μ_+, μ_-, μ_0) is referred to as the inertia of A and will be designated by $\mathcal{I}n(A)$. If F is a matrix valued function, then $F^{\#}(\lambda) = F(1/\lambda^*)^*$.

An $m \times m$ signature matrix J is a constant matrix with $J = J^*$ and $JJ^* = I_m$. In this paper we take $m = 2p$ and deal mostly with the specific signature matrices

$$J_0 = \begin{bmatrix} I_p & 0 \\ 0 & -I_p \end{bmatrix} \quad \text{and} \quad J_1 = \begin{bmatrix} 0 & I_p \\ I_p & 0 \end{bmatrix} .$$

$H^2_{j\times k}$ [resp. $L^2_{j\times k}$] denotes the space of $j \times k$ matrix valued functions with entries in the usual Hardy space $H^2(\mathbb{D})$ [resp. $L^2(\mathbb{T})$] and H^2_j [resp. L^2_j] is short for $H^2_{j\times 1}$ [resp. $L^2_{j\times 1}$]; $\underline{\underline{p}}$ designates the orthogonal projection of L^2_j onto H^2_j.

The symbol $<\ ,\ >$ will be used for the standard inner product:

$$< f, g >= \frac{1}{2\pi} \int_0^{2\pi} g(e^{i\theta})^* f(e^{i\theta}) d\theta ,$$

when f and g belong to $L^2_k(\mathbb{T})$. We shall also use this notation from time to time for arbitrary size matrix valued functions with square summable entries and not just for column vector valued functions. The same comment applies to $\underline{\underline{p}}$.

The notations

$$\rho_\omega(\lambda) = 1 - \lambda\omega^*$$

and

$$(R_\omega f)(\lambda) = \frac{f(\lambda) - f(\omega)}{\lambda - \omega}$$

(for the generalized backwards shift) will also be used.

2. REPRODUCING KERNEL PONTRYAGIN SPACES

In this section we shall give a quick introduction to a theory of structured reproducing kernel Pontryagin spaces. For ease of exposition we restrict ourselves to finite dimensional spaces because this will be adequate for the needs of this paper. A more complete analysis may be found in Alpay-Dym [AD], from which the present discussion is adapted. The theory itself is an extension of a theory of structured reproducing kernel Hilbert spaces which originates with de Branges [dB] to the setting of Pontryagin spaces.

THEOREM 2.1. *Let $f_0, \ldots, f_n$ be a set of $m \times p$ matrix valued functions which belong to $H^2_{m\times p}$ of the disc $\mathbb{D}$, let J be an $m \times m$ signature matrix and suppose that the $(n+1) \times (n+1)$ block matrix $\mathbb{P}$ with ij block entry (of size $p \times p$)*

$$\mathbb{P}_{ij} = < Jf_j, f_i > , \qquad i,j = 0, \ldots, n ,$$

is invertible. Then the space

$$\mathcal{M} = \text{the span of the columns of } \{f_0, \ldots, f_n\}$$

endowed with the J inner product is an $(n+1)p$ dimensional reproducing kernel Pontryagin space with reproducing kernel

$$K_\omega(\lambda) = \sum_{i,j=0}^{n} f_i(\lambda)(\mathbb{P}^{-1})_{ij} f_j(\omega)^* , \tag{2.1}$$

for every choice of ω and λ in $\mathbb{D}$.

The Theorem is easily verified by direct computation. You have only to recall that the assertion that $\mathcal{M}$ is a reproducing kernel space with respect to the J inner product means that, for every choice of $\omega \in \mathbb{D}$, $v \in \mathbb{C}^m$ and $f \in \mathcal{M}$,

(1) $K_\omega v \in \mathcal{M}$, and

(2) $< Jf, K_\omega v >= v^* f(\omega)$.

Moreover, it is readily checked that

$$K_\alpha(\beta)^* = K_\beta(\alpha)$$

and that there is only one such reproducing kernel.

THEOREM 2.2. *If $\mathcal{M}$ and $\mathbb{P}$ are as in Theorem 2.1 and if also $\mathcal{M}$ is R_0 invariant, then there exists a rational $m \times m$ matrix valued function U which is analytic in $\bar{\mathbb{D}}$ and J unitary on $\mathbb{T}$ such that the reproducing kernel for $\mathcal{M}$ can be written in the form*

$$K_\omega(\lambda) = \frac{J - U(\lambda) J U(\omega)^*}{\rho_\omega(\lambda)} . \tag{2.2}$$

The function U is unique up to a constant J unitary factor on the right and may in fact be specified by the recipe

$$U(\lambda) = I_m - \rho_\beta(\lambda) \sum_{i,j=0}^{n} f_i(\lambda)(\mathbb{P}^{-1})_{ij} f_j(\beta)^* J \tag{2.3}$$

for any point $\beta \in \mathbb{T}$. (Different choices of $\beta \in \mathbb{T}$ serve only to change the constant J unitary multiplier.) Moreover, the number of negative squares of the kernel $K_\omega(\lambda)$ is equal to the number of negative eigenvalues of $\mathbb{P}$.

PROOF. This is an immediate consequence of Theorem 6.12 of [AD], albeit in block form, since the presumed R_0 invariance forces $\mathcal{M}$ to be R_α invariant for every point $\alpha \in \mathbb{D}$. ∎

COROLLARY. *If $\mathcal{M}, \mathbb{P}$ and U are as in Theorem 2.2, then*

$$\frac{J - U(\lambda)JU(\omega)^*}{\rho_\omega(\lambda)} = \sum_{i,j=0}^{n} f_i(\lambda)(\mathbb{P}^{-1})_{ij} f_j(\omega)^* \ . \tag{2.4}$$

PROOF. The right hand sides of (2.1) and (2.2) must match, since the reproducing kernel is unique. ∎

The general theory sketched above will now be applied to a space based on the span of the columns of the $m \times p$ matrix polynomials

$$\begin{aligned} f_0(\lambda) &= v_0 \\ f_j(\lambda) &= \lambda^j v_0 + \lambda^{j-1} v_1 + \ldots + v_j \ , \qquad j = 1, \ldots, n \ , \end{aligned} \tag{2.5}$$

which are defined in terms of the $m \times p$ constant matrices

$$v_0 = \begin{bmatrix} I_p \\ h_0 \end{bmatrix} , \qquad v_j = \begin{bmatrix} 0 \\ 2h_j^* \end{bmatrix} , \qquad j = 1, \ldots, n \ .$$

Notice that in the present setting $m = 2p$.

THEOREM 2.3. *If H_n is invertible, then the span $\mathcal{M}$ of the columns of the matrix polynomials f_j, $j = 0, \ldots, n$, defined by (2.5) is an $(n+1)p$ dimensional reproducing kernel Pontryagin space with respect to the J_1 inner product. Moreover, its reproducing kernel is of the form (2.2), where U is uniquely specified up to a right constant J_1 unitary factor by the formula*

$$U(\lambda) = I_m - \frac{\rho_\alpha(\lambda)}{2} \sum_{i,j=0}^{n} f_i(\lambda) \gamma_{ij}^{(n)} f_j(\alpha)^* J_1 \tag{2.6}$$

in which α is any point on $\mathbb{T}$. Moreover,

$$\det U(\lambda) = c\lambda^t \tag{2.7}$$

for some constant c with $|c| = 1$ and some integer $t \geq 0$.

PROOF. It is readily checked that

$$< J_1 f_j, f_i > = 2h_{i-j} \ , \qquad i,j = 0, \ldots, n \ ,$$

and hence that, in the notation of the preceding two theorems, $\mathbb{P} = 2H_n$ is invertible. The evaluation

$$R_0 f_j = f_{j-1} , \qquad j = 1, \ldots, n ,$$

guarantees that $\mathcal{M}$ is R_0 invariant. The desired conclusion, apart from (2.7), is thus immediate from Theorem 2.2.

Finally, it follows from (2.2) that

$$J_1 - U(\lambda)^* J_1 U(\lambda) = 0$$

for every point $\lambda \in \mathbb{T}$ and hence by analytic continuation, that

$$U^{\#}(\lambda) J_1 U(\lambda) = J_1$$

for every nonzero $\lambda \in \mathbb{C}$. But this clearly implies that U is invertible for every nonzero $\lambda \in \mathbb{C}$ and hence, since U is a matrix polynomial, that its determinant, which is a scalar polynomial, is of the indicated form. ∎

3. LINEAR FRACTIONAL TRANSFORMATIONS

For any $2p \times 2p$ matrix valued function

$$U = \begin{bmatrix} U_{11} & U_{12} \\ U_{21} & U_{22} \end{bmatrix}$$

with $p \times p$ blocks U_{ij}, $i, j = 1, \ldots, n$, and any $p \times p$ matrix valued function G we define the linear fractional transformation

$$Z_U[G] = -(U_{21} - U_{22}G)(U_{11} - U_{12}G)^{-1} \tag{3.1}$$

for those points ω at which the indicated inverse exists.

LEMMA 3.1. *If U is a constant $2p \times 2p$ J_1 contractive matrix and G is a constant $p \times p$ matrix with $G + G^* > 0$, then*

$$U_{11} - U_{12}G \quad \textit{and} \quad U_{21} - U_{22}G$$

are invertible.

PROOF. Let $X = U_{11} - U_{12}G$ and $Y = U_{22}G - U_{21}$. Then the inequality

$$[I_p \quad -G^*]\{J_1 - U^* J_1 U\}[I_p \quad -G^*]^* \geq 0$$

implies that

$$X^*Y + Y^*X \geq G + G^* .$$

Thus, if $X\xi = 0$ for some vector $\xi \in \mathbb{C}^p$, then

$$0 \geq \xi^*(G + G^*)\xi$$

and hence $\xi = 0$. This proves that X is invertible. A similar argument disposes of Y. ∎

LEMMA 3.2. *If U and G are as in Lemma 3.1 and if also U is invertible with $W = U^{-1}$, then*

$$W_{22} + GW_{12} \quad \textit{and} \quad W_{21} + GW_{11}$$

are invertible.

PROOF. Since W is J_1 expansive,

$$[G \;\; I_p]\{WJ_1W^* - J_1\}[G \;\; I_p]^* \geq 0$$

and the rest follows much as in the proof of Lemma 3.1. ∎

LEMMA 3.3. *If U and G are as in Lemma 3.1, then $Z_U[G]$ is well defined and*

$$Z_U[G] + Z_U[G]^* > 0 \; .$$

If also U is invertible with $U^{-1} = W$, then

$$Z_U[G] = (W_{22} + GW_{12})^{-1}(W_{21} + GW_{11}) \; . \tag{3.2}$$

PROOF. The fact that the right hand side of (3.1) [resp. (3.2)] is well defined is immediate from Lemma 3.1 [resp. Lemma 3.2]. The remaining two assertions are straightforward calculations. ∎

LEMMA 3.4. *If U and V are constant $2p \times 2p$ J_1 contractive matrices and G is a constant $p \times p$ matrix with $G + G^* > 0$, then*

$$Z_{UV}[G] = Z_U[Z_V[G]] \; . \tag{3.3}$$

If U is J_1 unitary, then $Z_U[G]$ defines a one to one mapping of the set of constant $p \times p$ matrices G with $G + G^* > 0$ onto themselves.

PROOF. In view of Lemmas 3.1 and 3.3 all the indicated linear fractional transformations are well defined. The verification of the identity (3.3) is straightforward. Finally, if U is J_1 unitary, then so is $W = J_1U^*J_1 = U^{-1}$. Therefore, by (3.3),

$$Z_U[Z_W[G]] = Z_{UW}[G] = G \; ,$$

which serves to prove that Z_U is onto, and

$$Z_W[Z_U[G]] = Z_{WU}[G] = G \; ,$$

which, in turn, serves to establish the fact that Z_U is one to one. ∎

4. SOME USEFUL IDENTITIES

From now on we assume that the block Toeplitz matrix H_n is invertible, that $\mathcal{M}$ and U are defined as in Theorem 2.3, that

$$\Psi(\lambda) = h_0 + 2\sum_{j=1}^{n} h_j\lambda^j \tag{4.1}$$

and let $\mathcal{M}_+$ be the Pontryagin space of $p \times 1$ vector polynomials of degree less than or equal to n, endowed with the indefinite inner product

$$< u, v >_{\mathcal{M}_+} = < \frac{\Psi + \Psi^*}{2} u, v > \ .$$

Then, it is readily checked that

$$f_j\xi = \begin{bmatrix} \lambda^j\xi \\ \underline{\underline{p}}\Psi^*\lambda^j\xi \end{bmatrix}$$

and hence that (as a set)

$$\mathcal{M} = \left\{ \begin{bmatrix} u \\ \underline{\underline{p}}\Psi^* u \end{bmatrix} : \ u \in \mathcal{M}_+ \right\} ,$$

and furthermore, that if

$$f = \begin{bmatrix} u \\ \underline{\underline{p}}\psi^* u \end{bmatrix} , \quad u \in \mathcal{M}_+ ,$$

then

$$\begin{aligned} < f, f >_{\mathcal{M}} &= < J_1 f, f > \\ &= < u, \underline{\underline{p}}\psi^* u > + < \underline{\underline{p}}\psi^* u, u > \\ &= 2 < u, u >_{\mathcal{M}_+} \ . \end{aligned}$$

Thus the mapping

$$f \longrightarrow \sqrt{2}[I_p \ \ 0] f$$

is clearly an isometric isomorphism of $\mathcal{M}$ onto $\mathcal{M}_+$ and serves to identify the reproducing kernel of $\mathcal{M}_+$ as

$$\Lambda_\omega(\lambda) = 2[I_p \ \ 0] \left\{ \frac{J_1 - U(\lambda)J_1U(\omega)^*}{\rho_\omega(\lambda)} \right\} [I_p \ \ 0]^* \tag{4.2}$$

for all points λ and ω in $\mathbb{D}$.

LEMMA 4.1. *If H_n is invertible and if U is defined by (2.6), then the reproducing kernel for $\mathcal{M}_+$, $\Lambda_\omega(\lambda)$, admits the following three representations:*

$$\Lambda_\omega(\lambda) = \sum_{i,j=0}^{n} \lambda^i \gamma_{ij}^{(n)} \omega^{*j} \tag{4.3}$$

$$\Lambda_\omega(\lambda) = -2\frac{\{U_{11}(\lambda)U_{12}(\omega)^* + U_{12}(\lambda)U_{11}(\omega)^*\}}{\rho_\omega(\lambda)} \tag{4.4}$$

$$\begin{aligned}\Lambda_\omega(\lambda) = \Big(&\{U_{11}(\lambda) - U_{12}(\lambda)G\}\{(G+G^*)/2\}^{-1}\{U_{11}(\omega) - U_{12}(\omega)G\}^* \\ &-\{U_{11}(\lambda) + U_{12}(\lambda)G^*\}\{(G+G^*)/2\})^{-1}\{U_{11}(\omega) + U_{12}(\omega)G^*\}^*\Big)/\rho_\omega(\lambda)\end{aligned} \tag{4.5}$$

for any constant $p \times p$ matrix G with $G + G^*$ invertible.

PROOF. Formula (4.3) is immediate from (4.2), (2.4) and the fact that $\mathbb{P}^{-1} = \Gamma_n/2$. Formula (4.4) is a direct evaluation of (4.2) while (4.5) comes by reexpressing (4.4) as

$$\Lambda_\omega(\lambda) = -2\frac{[U_{11}(\lambda) \;\; U_{12}(\lambda)]J_1[U_{11}(\omega) \;\; U_{12}(\omega)]^*}{\rho_\omega(\lambda)}$$

upon invoking the simple but remarkably useful identity

$$J_1 = \begin{bmatrix} I_p & I_p \\ G^* & -G \end{bmatrix} \begin{bmatrix} (G+G^*)^{-1} & 0 \\ 0 & -(G+G^*)^{-1} \end{bmatrix} \begin{bmatrix} I_p & G \\ I_p & -G^* \end{bmatrix}, \tag{4.6}$$

which is valid for any $p \times p$ matrix G with invertible real part $G + G^*$. The verification of (4.6) is straightforward if you believe that

$$G(G+G^*)^{-1}G^* = G^*(G+G^*)^{-1}G\,. \tag{4.7}$$

But that in turn is easily justified upon expressing $G = E + iF$, in terms of its real part E and its imaginary part F. ∎

LEMMA 4.2. *If H_n and H_{n-1} are invertible, then*

(1) $\gamma_{00}^{(n)}$ and $\gamma_{nn}^{(n)}$ are invertible,

(2) $\mu_\pm(H_n) = \mu_\pm(H_{n-1}) + \mu_\pm(\gamma_{00}^{(n)})$,

(3) $\mu_\pm(H_n) = \mu_\pm(H_{n-1}) + \mu_\pm(\gamma_{nn}^{(n)})$, and

(4) $\mu_\pm(\gamma_{00}^{(n)}) = \mu_\pm(\gamma_{nn}^{(n)})$.

PROOF. Let $x_i = \gamma_{i0}^{(n)}$, $i = 1, \ldots, n$, $x = x_0$, $\beta = [h_{-1} \ldots h_{-n}]$ and

$$X = \begin{bmatrix} x_1 \\ \vdots \\ x_n \end{bmatrix}.$$

Then it follows readily from the block matrix identity

$$\begin{bmatrix} h_0 & \beta \\ \beta^* & H_{n-1} \end{bmatrix} \begin{bmatrix} x \\ X \end{bmatrix} = \begin{bmatrix} I_p \\ 0 \end{bmatrix}$$

that

$$(h_0 - \beta\Gamma_{n-1}\beta^*)\gamma_{00}^{(n)} = I_p \ .$$

Therefore $\gamma_{00}^{(n)}$ is invertible and (2) in turn is an immediate consequence of Sylvester's law of inertia (see e.g. Lancaster and Tismenetsky [LT]) applied to the identity

$$H_n = \begin{bmatrix} I_p & \beta\Gamma_{n-1} \\ 0 & I_{np} \end{bmatrix} \begin{bmatrix} \{\gamma_{00}^{(n)}\}^{-1} & 0 \\ 0 & H_{n-1} \end{bmatrix} \begin{bmatrix} I_p & 0 \\ \Gamma_{n-1}\beta^* & I_{np} \end{bmatrix} .$$

The asserted invertibility of $\gamma_{nn}^{(n)}$ and (3) follow in much the same way from the identity

$$\begin{bmatrix} H_{n-1} & \gamma^* \\ \gamma & h_0 \end{bmatrix} \begin{bmatrix} Y \\ y \end{bmatrix} = \begin{bmatrix} 0 \\ I_p \end{bmatrix}$$

wherein $\gamma = [h_n \dots h_1]$ and $\begin{bmatrix} Y \\ y \end{bmatrix}$ is the last block column of Γ_n. The rest is plain. ∎

LEMMA 4.3. *If H_n is invertible and if U is defined by (2.6), then*

$$A_n(\lambda) = \sum_{i=0}^{n} \lambda^i \gamma_{i0}^{(n)} = -2\{U_{11}(\lambda)U_{12}(0)^* + U_{12}(\lambda)U_{11}(0)^*\} \ . \tag{4.8}$$

If also H_{n-1} is invertible, then $U_{11}(0)$ and $U_{12}(0)$ are invertible and, if

$$G_o = -\{U_{12}(0)^{-1}U_{11}(0)\}^* \ , \tag{4.9}$$

then

$$G_o + G_o^* = \{U_{12}(0)\}^{-1}\gamma_{00}^{(n)}\{U_{12}(0)^*\}^{-1}/2 \ . \tag{4.10}$$

PROOF. Formula (4.8) is immediate from formulas (4.3) and (4.4) for $\Lambda_0(\lambda)$. Then, since $\gamma_{00}^{(n)}$ is invertible by Lemma 4.2 when H_{n-1} is also invertible, it is readily checked that the null spaces of $U_{11}(0)$ and $U_{12}(0)$ are both zero, and hence that these matrices are invertible. Finally, (4.10) is a straightforward calculation. ∎

LEMMA 4.4. *If H_n and H_{n-1} are invertible and if U and G_o are defined by (2.6) and (4.9) respectively, then*

$$A_n(\lambda) = -2\{U_{11}(\lambda) - U_{12}(\lambda)G_o\}U_{12}(0)^* \tag{4.11}$$

is invertible in a neighborhood of zero. If H_n is positive definite, then $A_n(\lambda)$ is invertible in all of $\mathbb{D}$.

PROOF. Formula (4.11) is immediate from (4.8) and (4.9). $A_n(\lambda)$ is invertible in a neighborhood of zero by Lemma 4.2. If $H_n > 0$, U is J_1 contractive in $\bar{\mathbb{D}}$ by (2.4), and $G_o + G_o^* > 0$ by (4.10). Therefore $U_{11}(\lambda) - U_{12}(\lambda)G_0$ is invertible in $\bar{\mathbb{D}}$ by Lemma 3.1. ∎

LEMMA 4.5. *If H_n is invertible and if U is defined as in (2.6), then*

$$\lim_{\lambda\to\infty} \frac{U_{11}(\lambda)}{\lambda^{n+1}} = \left\{\frac{\alpha C_n^o(\alpha)}{2}\right\}^* \tag{4.12}$$

and

$$\lim_{\lambda\to\infty} \frac{U_{12}(\lambda)}{\lambda^{n+1}} = \left\{\frac{\alpha C_n(\alpha)}{2}\right\}^* , \tag{4.13}$$

where $\alpha \in \mathbb{T}$ is the point which intervenes in the definition of U.

PROOF. Since the $p \times p$ matrix polynomial $p_j(\lambda)$ which is defined in (1.5) is equal to the bottom block of the $2p \times p$ matrix polynomial $f_j(\lambda)$ which is defined in (2.5), it follows readily from (2.6) that

$$2U_{11}(\lambda) = 2I_p - \rho_\alpha(\lambda) \sum_{i,j=0}^{n} \lambda^i \gamma_{ij}^{(n)} p_j(\alpha)^*$$

and

$$2U_{12}(\lambda) = -\rho_\alpha(\lambda) \sum_{i,j=0}^{n} \lambda^i \gamma_{ij}^{(n)} \alpha^{*j} .$$

The rest is a straightforward calculation. ∎

LEMMA 4.6. *If H_n is invertible and if U is defined by (2.6), then*

$$\begin{aligned}\lambda C_n(\lambda) &= \lambda \sum_{i=0}^{n} \lambda^i \gamma_{in}^{(n)} \\ &= \alpha\{U_{11}(\lambda)C_n(\alpha) + U_{12}(\lambda)C_n^o(\alpha)\} ,\end{aligned} \tag{4.14}$$

where $\alpha \in \mathbb{T}$ is the point which intervenes in the definition of U. If also H_{n-1} is invertible, then $C_n(\alpha)$ is invertible,

$$C_n^o(\alpha) = G_o^* C_n(\alpha) , \tag{4.15}$$

$$\lambda C_n(\lambda) = \{U_{11}(\lambda) + U_{12}(\lambda)G_o^*\}\alpha C_n(\alpha) , \tag{4.16}$$

$$2\gamma_{nn}^{(n)} = C_n(\alpha)^*\{G_o + G_o^*\}C_n(\alpha) \tag{4.17}$$

and the matrix polynomial $C_n(\lambda)$ is invertible in a neighborhood of infinity.

If H_n is positive definite, then $C_n(\lambda)$ is invertible for every point $\lambda \in \bar{\mathbb{E}}$.

PROOF. It follows readily from (4.3) that

$$\lambda C_n(\lambda) = \lim_{\omega \uparrow \infty} \left\{ -\frac{\rho_\omega(\lambda)\Lambda_\omega(\lambda)}{\omega^{*n+1}} \right\} .$$

Formula (4.14) is then obtained by invoking (4.4) and the evaluations of Lemma 4.5 in the preceding relationship.

If H_{n-1} is also invertible, then G_o is well defined by (4.9) and (4.15) drops out easily from (4.14) upon setting $\lambda = 0$. Formula (4.16) is then immediate from (4.14) and (4.15) and leads easily to the auxiliary identity

$$2\gamma_{nn}^{(n)} = \{C_n^o(\alpha)^* + C_n(\alpha)^* G_o^*\} C_n(\alpha) , \tag{4.18}$$

upon dividing by λ^{n+1}, letting $\lambda \uparrow \infty$ and invoking Lemma 4.5 once again. But this in turn yields (4.17) and serves also to justify the asserted invertibility of $C_n(\alpha)$, since $\gamma_{nn}^{(n)}$ is invertible by Lemma 4.2. Moreover, $C_n(\lambda)$ is clearly invertible in a neighborhood of infinity because its top coefficient is equal to $\gamma_{nn}^{(n)}$.

The fact that $C_n(\lambda)$ is invertible for every point $\lambda \in \bar{\mathbb{E}}$ when $H_n > 0$ will be established in the next lemma. ∎

LEMMA 4.7. *If H_n and H_{n-1} are invertible, then the reproducing kernel $\Lambda_\omega(\lambda)$ for $\mathcal{M}_+$ is equal to*

$$\sum_{i,j=0}^{n} \lambda^i \gamma_{ij}^{(n)} \omega^{*j} = \frac{A_n(\lambda)\{\gamma_{00}^{(n)}\}^{-1} A_n(\omega)^* - \lambda\omega^* C_n(\lambda)\{\gamma_{nn}^{(n)}\}^{-1} C_n(\omega)^*}{\rho_\omega(\lambda)} . \tag{4.19}$$

Moreover, $A_n(\lambda)$ and

$$\lambda^n C_n^{\#}(\lambda) = \sum_{i=0}^{n} \gamma_{ni}^{(n)} \lambda^{n-i} \tag{4.20}$$

are invertible in a neighborhood of zero.

If H_n is positive definite, then $A_n(\lambda)$ and $\lambda^n C_n^{\#}(\lambda)$ are invertible at every point $\lambda \in \bar{\mathbb{D}}$.

PROOF. Formula (4.19) is obtained by choosing $G = G_o$ in (4.5) and then invoking (4.10), (4.11), (4.16) and (4.17). The asserted invertibility of A_n and $\lambda^n C_n^{\#}$ near zero is immediate from Lemmas 4.4 and 4.6 and the fact that $\{\gamma_{ij}^{(n)}\}^* = \gamma_{ji}^{(n)}$.

Finally, if $H_n > 0$, then A_n is invertible in $\bar{\mathbb{D}}$ by Lemma 4.4, whereas the invertibility of C_n in $\mathbb{E}$ is easily deduced from (4.19) (upon taking advantage of already established properties of A_n the fact that $\xi^*\Lambda_\omega(\omega)\xi > 0$ for every point $\omega \in \mathbb{C}$ and nonzero vector $\xi \in \mathbb{C}^p$, since $\Gamma_n > 0$). ∎

We remark that (4.11) is equivalent to a well known formula of Gohberg and Heinig [GH] which serves to express Γ_n in terms of the entries in its first and last columns only:

$$\Gamma_n = \delta_n \begin{bmatrix} a_0 \\ \vdots \\ a_n \end{bmatrix} d_n(a_0^{-1})\delta_n \begin{bmatrix} a_0 \\ \vdots \\ a_n \end{bmatrix}^* - \delta_n \begin{bmatrix} c_0 \\ \vdots \\ c_n \end{bmatrix} d_n(c_{n+1}^{-1})\delta_n \begin{bmatrix} c_0 \\ \vdots \\ c_n \end{bmatrix}^* , \tag{4.21}$$

in which

$$a_i = \gamma_{i0}^{(n)}, \quad i = 0, \ldots, n \ , \quad \text{and} \quad c_0 = 0, \quad c_{i+1} = \gamma_{in}^{(n)}, \quad i = 0, \ldots, n \ ,$$

designate the coefficients of $A_n(\lambda)$ and $\lambda C_n(\lambda)$, respectively, $\delta_n[\ ]$ denotes the $(n+1) \times (n+1)$ block lower triangular Toeplitz matrix with indicated first block column and $d_n(\ )$ designates the $(n+1) \times (n+1)$ block diagonal Toeplitz matrix with the indicated entry.

It is perhaps well to emphasize that, as the present derivation indicates, this is but one of many such possible identities, corresponding to different choices of G in (4.5).

5. POLYNOMIALS OF THE SECOND KIND

Our next objective is to obtain analogues of the main results in Section 4 for U_{21} and U_{22} in terms of the polynomials of the second kind

$$A_n^o(\lambda) = \sum_{i=0}^{n} \lambda^i (L_n\Gamma_n)_{io} \quad \text{and} \quad C_n^o(\lambda) = \sum_{i=0}^{n} \lambda^i (L_n^*\Gamma_n)_{in} \ , \tag{5.1}$$

in which

$$L_n = \begin{bmatrix} h_0 & 0 & \ldots & 0 \\ 2h_1 & h_0 & \ldots & 0 \\ \vdots & & & \vdots \\ 2h_n & 2h_{n-1} & \ldots & h_0 \end{bmatrix} .$$

is the block lower triangular Toeplitz matric based on the coefficients of Ψ. The recipes of (5.1) are of course equivalent to those of (1.4), wherein A_n^o and C_n^o are expressed in terms of the polynomials

$$\begin{aligned} p_i(\lambda) &= [0 \quad I_p] f_i(\lambda) \\ &= \underline{\underline{p}}\Psi^*\lambda^i \\ &= \begin{cases} h_0^* = h_0 & \text{if } i = 0 \\ h_0^*\lambda^i + 2h_1^*\lambda^{i-1} + \ldots + 2h_i^* & \text{if } i = 1, \ldots, n \ . \end{cases} \end{aligned} \tag{5.2}$$

A proof of the equivalence of the two definitions for A_n^o is provided in the proof of Lemma 5.4. The proof of the equivalence of the two definitions of C_n^o is even easier, especially with the help of (5.9). For other characterizations of these polynomials see Lemmas 7.4 and 7.5.

We continue to presume that H_n is invertible and take $\mathcal{M}$ and U as in Theorem 2.3. The kernel

$$\Lambda_\omega^\bullet(\lambda) = 2[0 \quad I_p]\left\{\frac{J_1 - U(\lambda)J_1U(\omega)^*}{\rho_\omega(\lambda)}\right\}[0 \quad I_p]^* \tag{5.3}$$

plays a useful role.

LEMMA 5.1. *If H_n is invertible, then*

$$L_n^*\Gamma_n L_n = L_n\Gamma_n L_n^* \ . \tag{5.4}$$

PROOF. Let us write

$$L_n = X_n + iY_n$$

where X_n and Y_n are selfadjoint. Then

$$X_n = \frac{L_n + L_n^*}{2} = H_n$$

is invertible by assumption and so

$$\begin{aligned} L_n^*\Gamma_n L_n &= (X_n - iY_n)X_n^{-1}(X_n + iY_n) \\ &= X_n + Y_n X_n^{-1} Y_n \\ &= (X_n + iY_n)X_n^{-1}(X_n - iY_n) \\ &= L_n\Gamma_n L_n^* \ . \end{aligned}$$

∎

LEMMA 5.2. *If H_n is invertible and if U is defined by (2.6), then the kernel $\Lambda_\omega^\bullet(\lambda)$ admits the following three representations:*

$$\Lambda_\omega^\bullet(\lambda) = \sum_{i,j=0}^{n} \lambda^i \dot\gamma_{ij}^{(n)} \omega^{*j} \ , \tag{5.5}$$

where

$$\dot\gamma_{ij}^{(n)} = (L_n^*\Gamma_n L_n)_{ij} \tag{5.6}$$

denotes the ij block of $L_n^*\Gamma_n L_n$,

$$\Lambda_\omega^\bullet(\lambda) = -2\{U_{21}(\lambda)U_{22}(\omega)^* + U_{22}(\lambda)U_{21}(\omega)^*\}/\rho_\omega(\lambda) \tag{5.7}$$

and

$$\begin{aligned} \Lambda_\omega^\bullet(\lambda) = \Big(&\{U_{21}(\lambda) - U_{22}(\lambda)G\}\{(G+G^*)/2)\}^{-1}\{U_{21}(\omega) - U_{22}(\omega)G\}^* \\ &-\{U_{21}(\lambda) + U_{22}(\lambda)G^*\}\{(G+G^*)/2)\}^{-1}\{U_{21}(\omega) + U_{22}(\omega)G^*\}^*\Big)/\rho_\omega(\lambda) \end{aligned} \tag{5.8}$$

for any constant $p \times p$ matrix G with $G + G^*$ invertible.

PROOF. The identity

$$\Lambda_\omega^\bullet(\lambda) = \sum_{i,j=0}^{n} p_i(\lambda)\gamma_{ij}^{(n)} p_j(\omega)^*$$

with $p_i(\lambda)$ as in (5.2) is established much as in the proof of (4.3). The trick now is to observe that

$$p_i(\lambda) = \sum_{s=0}^{i} (L_n^*)_{si}\lambda^s \quad \text{and} \quad p_j(\omega)^* = \sum_{t=0}^{j} (L_n)_{jt}\omega^{*t} \tag{5.9}$$

and hence that

$$\begin{aligned}\Lambda_\omega^\bullet(\lambda) &= \sum_{i,j=0}^{n} \left\{ \sum_{s=0}^{i} (L_n^*)_{si}\lambda^s (\Gamma_n)_{ij} \sum_{t=0}^{j} (L_n)_{jt}\omega^{*t} \right\} \\ &= \sum_{s,t=0}^{n} \lambda^s \left\{ \sum_{i=s}^{n} (L_n^*)_{si} \sum_{j=t}^{n} (\Gamma_n)_{ij}(L_n)_{jt} \right\} \omega^{*t} ,\end{aligned}$$

which is readily seen to yield (5.5).

The last two formulas can be established by imitating the proofs of (4.4) and (4.5). ∎

LEMMA 5.3. *If H_n is invertible and if U is defined by (2.6), then*

$$U_{21}(0) = -h_0 U_{11}(0) \quad \textit{and} \quad U_{22}(0) = -h_0 U_{12}(0) \tag{5.10}$$

PROOF. This is a straightforward calculation based on (2.6) and the fact that

$$\sum_{i=0}^{n} h_{-i}\gamma_{ij}^{(n)} = \begin{cases} I_p & \text{for} \quad j = 0 \\ 0 & \text{for} \quad j = 1, \ldots, n \ . \end{cases}$$

It may also be obtained a little more elegantly from Lemma 7.1 and the observation that (5.10) is equivalent to the assertion that

$$v_0^* J_1 U(0) = 0 \ .$$ ∎

LEMMA 5.4. *If H_n is invertible and if U is defined by (2.6), then*

$$A_n^o(\lambda) = 2\{U_{21}(\lambda)U_{12}(0)^* + U_{22}(\lambda)U_{11}(0)^*\} \tag{5.11}$$

If also H_{n-1} and h_0 are invertible, then $A_n^o(\lambda)$ is invertible in a neighborhood of zero. If H_n is positive definite, then $A_n^o(\lambda)$ is invertible for every point $\lambda \in \mathbb{D}$.

PROOF. Upon matching the 21 block entries of formula (2.4) with $J = J_1$, f_i as in (2.5), U as in (2.6) and $\omega = 0$, it is readily seen that

$$\frac{1}{2}\sum_{i=0}^{n} p_i(\lambda)\gamma_{i0}^{(n)} = I_p - \{U_{21}(\lambda)U_{12}(0)^* + U_{22}(\lambda)U_{11}(0)^*\} .$$

But now, with the aid of (5.9), the sum on the left can be reexpressed as

$$\begin{aligned}\sum_{i=0}^{n} p_i(\lambda)\gamma_{i0}^{(n)} &= \sum_{i=0}^{n}\sum_{s=0}^{i} \lambda^s (L_n^*)_{si}\gamma_{i0}^{(n)} = \sum_{s=0}^{n} \lambda^s \sum_{i=s}^{n} (L_n^*)_{si}\gamma_{i0}^{(n)} \\ &= \sum_{s=0}^{n} \lambda^s (L_n^*\Gamma_n)_{s0} = \sum_{s=0}^{n} \lambda^s (2H_n\Gamma_n - L_n\Gamma_n)_{s0} = 2I_p - A_n^o(\lambda) .\end{aligned}$$

The desired formula drops out easily upon combining terms.

The asserted invertibility goes through much as in the proof of Lemma 4.4 except that h_0 intervenes because $A_n^o(0) = h_0 A_n(0)$ by Lemma 5.3. ∎

LEMMA 5.5. *If H_n is invertible and if U is defined by (2.6), then*

$$A_n(\lambda) = U_{11}(\lambda)A_n(\alpha) - U_{12}(\lambda)A_n^o(\alpha) , \tag{5.12}$$

$$-A_n^o(\lambda) = U_{21}(\lambda)A_n(\alpha) - U_{22}(\lambda)A_n^o(\alpha) , \tag{5.13}$$

$$\lambda C_n(\lambda) = \alpha\{U_{11}(\lambda)C_n(\alpha) + U_{12}(\lambda)C_n^o(\alpha)\} , \tag{5.14}$$

$$\lambda C_n^o(\lambda) = \alpha\{U_{21}(\lambda)C_n(\alpha) + U_{22}(\lambda)C_n^o(\alpha)\} , \tag{5.15}$$

$$2\gamma_{00}^{(n)} = A_n^o(\alpha)^* A_n(\alpha) + A_n(\alpha)^* A_n^o(\alpha) , \tag{5.16}$$

$$0 = A_n^o(\alpha)^* C_n(\alpha) - A_n(\alpha)^* C_n^o(\alpha) , \tag{5.17}$$

$$2\gamma_{nn}^{(n)} = C_n^o(\alpha)^* C_n(\alpha) + C_n(\alpha)^* C_n^o(\alpha) , \tag{5.18}$$

where $\alpha \in \mathbb{T}$ is the point which intervenes in the definition of U.

PROOF. Since $U(\alpha) = I_m$ it is easily seen from (4.8) and (5.11), respectively, that

$$A_n(\alpha) = -2U_{12}(0)^* \quad \text{and} \quad A_n^o(\alpha) = 2U_{11}(0)^* . \tag{5.19}$$

Formulas (5.12) and (5.13) are then immediate from (4.8) and (5.11), whereas (5.16) is just (5.12) evaluated at $\lambda = 0$, thanks to (5.19).

Next, it follows readily from the 21 block entry of formula (2.4), much as in the proof of the preceding lemma, that

$$\begin{aligned}\lambda C_n^o(\lambda) &= \lambda \sum_{i=0}^{n} p_i(\lambda)\gamma_{in}^{(n)} \\ &= \lim_{\omega^* \to \infty} 2\left\{\frac{U_{21}(\lambda)U_{12}(\omega)^* + U_{22}(\lambda)U_{11}(\omega)^*}{\omega^{*n+1}}\right\} .\end{aligned}$$

But, by Lemma 4.5, this yields (5.15). Formula (5.14) is recorded here for the sake of completeness. It has already been established in Lemma 4.6. Finally, (5.17) is immediate from (5.14) and (5.19) upon setting $\lambda = 0$, whereas (5.18) is immediate from (4.15) and (4.17). ∎

LEMMA 5.6. *If H_n and H_{n-1} are invertible and if U and G_o are defined by (2.6) and (4.9), respectively, then $A_n(\alpha)$ and $C_n(\alpha)$ are invertible and*

$$A_n^o(\alpha) = G_o A_n(\alpha) \tag{5.20}$$

$$A_n(\lambda) = \{U_{11}(\lambda) - U_{12}(\lambda)G_o\}A_n(\alpha) \ , \tag{5.21}$$

$$-A_n^o(\lambda) = \{U_{21}(\lambda) - U_{22}(\lambda)G_o\}A_n(\alpha) \ , \tag{5.22}$$

$$\lambda C_n(\lambda) = \{U_{11}(\lambda) + U_{12}(\lambda)G_o^*\}\alpha C_n(\alpha) \ , \tag{5.23}$$

$$\lambda C_n^o(\lambda) = \{U_{21}(\lambda) + U_{22}(\lambda)G_o^*\}\alpha C_n(\alpha) \ , \tag{5.24}$$

$$2\gamma_{00}^{(n)} = A_n(\alpha)^*(G_o + G_o^*)A_n(\alpha) \ , \tag{5.25}$$

$$2\gamma_{nn}^{(n)} = C_n(\alpha)^*(G_o + G_o^*)C_n(\alpha) \ , \tag{5.26}$$

where $\alpha \in \mathbb{T}$ is the point which intervenes in the definition of U.

If also h_0 is invertible, then $C_n^o(\lambda)$ is invertible in a neighborhood of infinity.

If H_n is positive definite, then $C_n^o(\lambda)$ is invertible for every point $\lambda \in \bar{\mathbb{E}}$.

PROOF. It follows readily from (5.16) and (5.18), respectively, that $A_n(\alpha)$ and $C_n(\alpha)$ are invertible (the invertibility of the latter is also available from Lemma 4.6). Formula (5.20) is thus immediate from (5.17) and (4.15), whereas (5.21)-(5.24) are immediate from (5.12)-(5.15), with the aid of (4.15) and (5.20). Similarly (5.25) and (5.26) are immediate from (5.16) and (5.18).

Next, $C_n^o(\lambda)$ is invertible in a neighborhood of infinity when h_0 is also invertible because its top coefficient is equal to $h_0\gamma_{nn}^{(n)}$.

Finally, the asserted invertibility for $H_n > 0$ is considered in the next lemma. ∎

LEMMA 5.7. *If H_n and H_{n-1} are invertible and if U is defined by (2.6), then*

$$\sum_{i,j=0}^{n} \lambda^i \gamma_{ij}^{(n)} \omega^{*j} = \frac{A_n^o(\lambda)\{\gamma_{00}^{(n)}\}^{-1}A_n^o(\omega)^* - \lambda\omega^* C_n^o(\lambda)\{\gamma_{nn}^{(n)}\}^{-1}C_n^o(\omega)^*}{\rho_\omega(\lambda)} \ . \tag{5.27}$$

If H_n is positive definite, then $A_n^o(\lambda)$ and $\lambda^n C_n^{o\#}(\lambda)$ are invertible for every point $\lambda \in \bar{\mathbb{D}}$.

PROOF. The formula is immediate from (5.5) and (5.8) upon fixing $G = G_o$ in the latter and taking advantage of (5.22) and (5.24)-(5.26).

The asserted invertibility of A_n^o is covered by Lemma 5.4, while the verification for $\lambda^n C_n^{o\#}$ follows from (5.27) and the fact that $L_n^*\Gamma_n L_n > 0$ as $H_n > 0$, just as in the proof of the analogous fact in Lemma 4.7. ∎

6. SOME MATRIX IDENTITIES

THEOREM 6.1. *If H_n is invertible, then the matrix polynomial*

$$\Theta_n(\lambda) = \begin{bmatrix} \lambda C_n(\lambda) & A_n(\lambda) \\ \lambda C_n^o(\lambda) & -A_n^o(\lambda) \end{bmatrix} \tag{6.1}$$

satisfies the identity

$$\Theta_n^\#(\lambda) J_1 \Theta_n(\lambda) = 2\Delta_n \tag{6.2}$$

for every nonzero point $\lambda \in \mathbb{C}$, where

$$\Delta_n = \begin{bmatrix} \gamma_{nn}^{(n)} & 0 \\ 0 & -\gamma_{00}^{(n)} \end{bmatrix} . \tag{6.3}$$

If both H_n and H_{n-1} are invertible, then also

$$\Theta_n(\lambda)\Delta_n^{-1}\Theta_n^\#(\lambda) = 2J_1 . \tag{6.4}$$

PROOF. If H_n is invertible and U is defined by (2.6), then, by Lemma 5.5,

$$\Theta_n(\lambda) = U(\lambda)\Theta_n(\alpha) . \tag{6.5}$$

Therefore,

$$\begin{aligned} \Theta_n^\#(\lambda) J_1 \Theta_n(\lambda) &= \Theta_n(\alpha)^* U^\#(\lambda) J_1 U(\lambda) \Theta_n(\alpha) \\ &= \Theta_n(\alpha)^* J_1 \Theta_n(\alpha) \\ &= 2\Delta_n , \end{aligned}$$

by formulas (5.16)-(5.18).

If both H_n and H_{n-1} are invertible, then Δ_n is invertible and therefore, by (6.2),

$$\begin{aligned} I_m &= \Theta_n^\#(\lambda) J_1 \Theta_n(\lambda)(2\Delta_n)^{-1} \\ &= J_1 \Theta_n(\lambda)(2\Delta_n)^{-1}\Theta_n^\#(\lambda) \end{aligned}$$

for every nonzero point $\lambda \in \mathbb{C}$. The rest is plain. ∎

COROLLARY 1. *If H_n is invertible, then*

$$\delta_n = 2^m \det\{\gamma_{00}^{(n)}\}\det\{\gamma_{nn}^{(n)}\} \geq 0$$

and

$$\det \Theta_n(\lambda) = \gamma\lambda^\nu , \quad \textit{with} \quad |\gamma| = \delta_n^{\frac{1}{2}} ,$$

for some choice of $\nu \geq 0$.

PROOF. Let

$$\varphi(\lambda) = \det \Theta_n(\lambda) .$$

Then it follows readily from (6.2) that

$$\varphi^{\#}(\lambda)\varphi(\lambda) = \delta_n \tag{6.6}$$

for every nonzero $\lambda \in \mathbb{C}$. In particular, $\delta_n \geq 0$ emerges by choosing $\lambda \in \mathbb{T}$. Since φ is a polynomial it can be expressed in the form

$$\varphi(\lambda) = \gamma(\lambda - \beta_1)\dots(\lambda - \beta_\nu)$$

with say $|\beta_1| \leq \dots \leq |\beta_\nu|$. But then (6.6) implies that

$$|\gamma|^2(1 - \beta_1^*\lambda)\dots(1 - \beta_\nu^*\lambda)(\lambda - \beta_1)\dots(\lambda - \beta_\nu) = \lambda^\nu \delta_n$$

and hence that the coefficients of $\lambda^{\nu+1}, \dots, \lambda^{2\nu}$ on the left must be equal to zero. Therefore, at least $\beta_1 = 0$ and the preceding formula reduces to

$$|\gamma|^2(1 - \beta_2^*\lambda)\dots(1 - \beta_\nu^*\lambda)(\lambda - \beta_2)\dots(\lambda - \beta_\nu) = \lambda^{\nu-1}\delta_n .$$

But now, upon iterating the argument, it is readily seen that $\beta_2 = \dots = \beta_\nu = 0$, as needed. ∎

COROLLARY 2. *If H_n and H_{n-1} are invertible, then $A_n(\lambda)$, $A_n^o(\lambda)$, $C_n(\lambda)$ and $C_n^o(\lambda)$ are invertible for every point $\lambda \in \mathbb{T}$.*

PROOF. It follows from the 11 blocks of (6.2) evaluated on $\mathbb{T}$:

$$C_n(\lambda)^* C_n^o(\lambda) + C_n^o(\lambda)^* C_n(\lambda) = 2\gamma_{nn}^{(n)} ,$$

that if $C_n(\lambda)\xi = 0$ for some $\lambda \in \mathbb{T}$ and $\xi \in \mathbb{C}^p$, then $\xi^*\gamma_{nn}^{(n)}\xi = 0$ and hence, since $\gamma_{nn}^{(n)}$ is invertible, that $\xi = 0$. Therefore the null space of the matrix $C_n(\lambda)$ is equal to zero on $\mathbb{T}$. This proves that $C_n(\lambda)$ is invertible for every point $\lambda \in \mathbb{T}$ as is $C_n^o(\lambda)$, by the very same argument. The invertibility of $A_n(\lambda)$ and $A_n^o(\lambda)$ is deduced in much the same way from the identity

$$A_n(\lambda)^* A_n^o(\lambda) + A_n^o(\lambda)^* A_n(\lambda) = 2\gamma_{00}^{(n)} ,$$

which is valid for every point $\lambda \in \mathbb{T}$, as follows by evaluating the 22 block of (6.2). ∎

THEOREM 6.2. *If H_n and H_{n-1} are invertible, then there exists a constant $m \times m$ matrix W_n such that*

$$\Delta_n^{-1} = 2W_n J_1 W_n^*$$

and hence, for every such W_n, $\Theta_n W_n$ is J_1 unitary on $\mathbb{T}$. If $\gamma_{00}^{(n)}$ is positive definite, then $\Delta_n^{-1}J_0$ is positive definite and it is possible to choose

$$W_n = (\Delta_n^{-1}J_0)^{\frac{1}{2}}\frac{1}{2}\begin{bmatrix} I_p & I_p \\ I_p & -I_p \end{bmatrix} ,$$

where the superscript $\frac{1}{2}$ designates the positive square root of the indicated matrix.

PROOF. If H_n and H_{n-1} are invertible, then, by Lemma 4.2, there exists a pair of $p \times p$ matrices X_n and Y_n and a $p \times p$ signature matrix j such that

$$\Delta_n^{-1} = \begin{bmatrix} X_n & 0 \\ 0 & Y_n \end{bmatrix} \begin{bmatrix} j & 0 \\ 0 & -j \end{bmatrix} \begin{bmatrix} X_n & 0 \\ 0 & Y_n \end{bmatrix}^* .$$

The existence of a W_n which meets the requisite identity is now clear since the center matrix on the right in the last formula is clearly unitarily equivalent to J_0 and hence also to J_1. The asserted J_1 unitary of $\Theta_n W_n$ is now immediate from (6.4). The rest is selfevident. ∎

7. INTERPOLATION

We shall say that a $p \times p$ matrix valued meromorphic function Φ is a Pade approximant or an interpolant of

$$\Psi(\lambda) = h_0 + 2h_1\lambda + \ldots + 2h_n\lambda^n$$

if Φ is analytic in a neighborhood of zero and

$$\Phi(\lambda) = \Psi(\lambda) + O(\lambda^{n+1}) \tag{7.1}$$

as $\lambda \to 0$. In this section we shall characterize such interpolants in terms of linear fractional transformations of more general classes of meromorphic functions. We shall assume throughout that at least H_n is invertible.

It is convenient to introduce the following subclasses of the class $\mathcal{M}_{p\times p}$ of $p \times p$ matrix valued functions which are meromorphic in $\mathbb{D}$ (though actually a neighborhood of zero would suffice):

$\mathcal{L}_1$:	$G \in \mathcal{M}_{p\times p}$	which are analytic in a neighborhood of zero,
$\mathcal{L}_2(k)$:	$F \in \mathcal{M}_{p\times p}$	which have a pole at zero and $\lambda^k F(\lambda)$ is invertible at zero,
$\mathcal{Q}_1(H_n)$:	$\Phi \in \mathcal{L}_1$	such that (7.1) holds,
$\mathcal{Q}_2(H_n;k)$:	$\Phi \in \mathcal{Q}_1(H_n)$	such that $(A_n^o - \Phi A_n)/\lambda^{n+k+1}$ is invertible at zero.

LEMMA 7.1. *If H_n is invertible and if $\mathcal{M}$, U, f_j and v_j are as in Theorem 2.3, then*

$$\lim_{\lambda\to 0} \left\{ \frac{v_0^* + \lambda v_1^* + \ldots + \lambda^n v_n^*}{\lambda^n} \right\} J_1 U(\lambda) = 0 \ . \tag{7.2}$$

PROOF. The fact that $\mathcal{M}$ is a reproducing kernel Pontryagin space clearly implies that

$$< J_1 f_t \xi, \frac{J_1 - UJ_1U(\omega)^*}{\rho_\omega} \chi >= \chi^* f_t(\omega)\xi$$

for $t = 0, \ldots, n$ and every choice of $\omega \in \mathbb{D}$, $\xi \in \mathbb{C}^p$ and $\chi \in \mathbb{C}^m$. Moreover, since U is a matrix polynomial, the inner product can be split into

$$< J_1 f_t \xi, \frac{J_1 \chi}{\rho_\omega} > - < J_1 f_t \xi, \frac{U J_1 U(\omega)^*}{\rho_\omega} \chi >= \chi^* f_t(\omega)\xi - < J_1 f_t \xi, \frac{U J_1 U(\omega)^*}{\rho_\omega} \chi > .$$

But this in turn implies that

$$< \frac{U J_1 U(\omega)}{\rho_\omega} \chi, \quad J_1 f_t \xi >= 0$$

and hence, since by (2.7) $U(\omega)$ is invertible for $\omega \neq 0$, that

$$\frac{1}{2\pi} \int_0^{2\pi} \frac{f_t(e^{i\theta})^* J_1 U(e^{i\theta})}{\rho_\omega(e^{i\theta})} d\theta = 0$$

for $t = 0, \ldots, n$ first for every nonzero $\omega \in \mathbb{D}$ and then, by passing to the limit, for $\omega = 0$ also. Thus, upon setting $\omega = 0$ and writing

$$U(\lambda) = \sum_{j=0}^{\infty} U_j \lambda^j$$

as a power series about zero with matrix coefficients U_j, it follows readily from Cauchy's formula that

$$\sum_{j=0}^{k} v_j^* J_1 U_{k-j} = 0 , \qquad k = 0, \ldots, n .$$

The desired conclusion now drops out easily upon identifying the last sum as the coefficient of λ^k in the product

$$(v_0^* + \lambda v_1^* + \ldots + \lambda^n v_n^*) J_1 U(\lambda) .$$ ∎

COROLLARY. *If H_n is invertible, then*

$$[\Psi(\lambda) \;\; I_p]\Theta_n(\lambda) = \lambda^{n+1}[P(\lambda) \;\; Q(\lambda)] , \tag{7.3}$$

where P and Q are matrix polynomials.

PROOF. (7.2) is clearly valid with Θ_n in place of U because of (6.5). The rest is selfevident since

$$[\Psi(\lambda) \;\; I_p] = (v_0^* + \lambda v_1^* + \ldots + \lambda^n v_n^*) J_1 .$$ ∎

LEMMA 7.2. *If H_n is invertible, if F and G belong to $\mathcal{L}_1$ and $A_n(\lambda)F(\lambda) + \lambda C_n(\lambda)G(\lambda)$ is invertible in a neighborhood of zero, then*

$$\Phi(\lambda) = \{A_n^o(\lambda)F(\lambda) - \lambda C_n^o(\lambda)G(\lambda)\}\{A_n(\lambda)F(\lambda) + \lambda C_n(\lambda)G(\lambda)\}^{-1} \tag{7.4}$$

belongs to $\mathcal{Q}_1(H_n)$.

PROOF. It follows readily from (7.3) that

$$[\Psi(\lambda)\ \ I_p]\Theta_n(\lambda)\begin{bmatrix}G(\lambda)\\F(\lambda)\end{bmatrix} = \lambda^{n+1}[P(\lambda)\ \ Q(\lambda)]\begin{bmatrix}G(\lambda)\\F(\lambda)\end{bmatrix}$$

and hence, upon carrying out the indicated multiplications, that

$$\Psi(\lambda C_n G + A_n F) + \lambda C_n^o G - A_n^o F = \lambda^{n+1}(PG + QF)\ .$$

Therefore,

$$\Psi - \Phi = \lambda^{n+1}(PG + QF)(A_n F + \lambda C_n G)^{-1}$$

and hence $\Phi \in \mathcal{Q}_1$. ∎

It is convenient to have a version of the last lemma available in terms of U also.

LEMMA 7.3. *If H_n is invertible and U is defined by (2.6), if $G \in \mathcal{L}_1$ and $U_{11} - U_{12}G$ is invertible at zero, then $Z_U[G] \in \mathcal{Q}_1(H_n)$.*

PROOF. Lemma 7.1 implies that

$$[\Psi\ \ I_p]U = \lambda^{n+1}[P_1\ \ Q_1]$$

for a pair of matrix polynomials P_1 and Q_1. The rest follows upon multiplying through on the right by $\begin{bmatrix}I_p\\-G\end{bmatrix}$ much as in the proof of the preceding lemma. ∎

COROLLARY. *If H_n and H_{n-1} are invertible and if U and G_o are given by (2.6) and (4.9), respectively, then*

$$Z_U[G_o] = A_n^o(\lambda)\{A_n(\lambda)\}^{-1} \tag{7.5}$$

is an interpolant.

PROOF. Lemma 7.3 is applicable because $U_{11}(\lambda) - U_{12}(\lambda)G_o$ is invertible at zero, thanks to Lemma 4.4. The rest is plain from (3.1) and Lemma 5.7. ∎

The preceding corollary implies that *if H_n and H_{n-1} are invertible, then*

$$A_n^o(\lambda) = \Psi(\lambda)A_n(\lambda) + O(\lambda^{n+1}) \tag{7.6}$$

in a neighborhood of zero.

Formula (7.6) is true more generally and in fact serves to characterize A_n^o.

LEMMA 7.4. *If H_n is invertible and if $P(\lambda) = \sum_{j=0}^n P_j\lambda^j$ is a $p \times p$ matrix polynomial of degree n, then*

$$P(\lambda) - \Psi(\lambda)A_n(\lambda) = O(\lambda^{n+1}) \tag{7.7}$$

in a neighborhood of zero if and only if $P = A_n^o$.

PROOF. The given relationship holds if and only if P_j is equal to the coefficient of λ^j in ΨA_n for $j = 0, \ldots, n$, or equivalently if and only if

$$\begin{bmatrix} P_0 \\ \vdots \\ P_n \end{bmatrix} = L_n \begin{bmatrix} \gamma_{00}^{(n)} \\ \vdots \\ \gamma_{n0}^{(n)} \end{bmatrix} .$$

The rest is plain from (5.1). ∎

There is an analogous characterization of C_n^o.

LEMMA 7.5. *If H_n is invertible, then*

$$C_n^o(\lambda) + \Psi(\lambda)C_n(\lambda) = 2\lambda^n I_p + O(\lambda^{n+1}) \tag{7.8}$$

in a neighborhood of zero. Moreover, if $P(\lambda) = \sum_{j=0}^n P_j \lambda^j$ is a $p \times p$ matrix polynomial of degree n, then

$$P(\lambda) + \Psi(\lambda)C_n(\lambda) = 2\lambda^n I_p + O(\lambda^{n+1}) . \tag{7.9}$$

in a neighborhood of zero if and only if

$$P(\lambda) = C_n^o(\lambda) .$$

PROOF. The coefficient of λ^j, $j = 0, \ldots, n$, on the left hand side of (7.9) is equal to

$$P_j + (L_n\Gamma_n)_{jn} = P_j + 2(H_n\Gamma_n)_{jn} - (L_n^*\Gamma_n)_{jn} .$$

Therefore, with the help of (5.1), the given claim is readily seen to be equivalent to

$$P(\lambda) - C_n^o(\lambda) = O(\lambda^{n+1}) .$$

The rest is plain. ∎

It is convenient to define the pair of linear fractional transformations

$$\tau_n[G] = \{A_n^o(\lambda) - \lambda C_n^o(\lambda)G(\lambda)\}\{A_n(\lambda) + \lambda C_n(\lambda)G(\lambda)\}^{-1} \tag{7.10}$$

and its inverse

$$\hat{\tau}_n[\Phi] = \{\lambda\Phi(\lambda)C_n(\lambda) + \lambda C_n^o(\lambda)\}^{-1}\{A_n^o(\lambda) - \Phi(\lambda)A_n(\lambda)\} \tag{7.11}$$

based on Θ_n for those $p \times p$ matrix valued meromorphic functions G and Φ, respectively, for which the expressions are meaningful.

THEOREM 7.1. *If H_n and H_{n-1} are invertible, then τ_n defines a one to one map of $\mathcal{L}_1$ onto $\mathcal{Q}_1(H_n)$, which is inverted by $\hat{\tau}_n$.*

PROOF. Lemma 7.2 guarantees that τ_n maps $\mathcal{L}_1$ into $\mathcal{Q}_1$. Moreover, if

$$\Phi = \tau_n[G] ,$$

then it is readily seen that

$$\left\{\frac{\Phi(\lambda)C_n(\lambda)+C_n^o(\lambda)}{\lambda^n}\right\}G(\lambda)=\frac{A_n^o(\lambda)-\Phi(\lambda)A_n(\lambda)}{\lambda^{n+1}}$$

and hence, since the term multiplying G is invertible in a neighborhood of zero by (7.8), that the mapping is one to one and that $G=\hat{\tau}_n[\Phi]$.

Next, for any $\Phi\in\mathcal{Q}_1$, it follows readily from (7.6) and (7.8) that $G(\lambda)=\hat{\tau}_n[\Phi]$ belongs to $\mathcal{L}_1$. Moreover, the development

$$\begin{aligned}A_n^o-\lambda C_n^oG &= A_n^o-\lambda(C_n^o+\Phi C_n-\Phi C_n)G\\ &= A_n^o-(A_n^o-\Phi A_n)+\lambda\Phi C_nG\\ &= \Phi(A_n+\lambda C_nG)\end{aligned}$$

clearly implies that $\Phi=\tau_n[G]$. ∎

The classes $\mathcal{L}_2(k)$ and $\mathcal{Q}_2(H_n;k)$ are similarly related by means of the linear fractional transformation

$$\sigma_n[F]=\{A_n^o(\lambda)F(\lambda)-\lambda C_n^o(\lambda)\}\{A_n(\lambda)F(\lambda)+\lambda C_n(\lambda)\}^{-1} \tag{7.12}$$

and its inverse

$$\hat{\sigma}_n[\Phi]=\{A_n^o(\lambda)-\Phi(\lambda)A_n(\lambda)\}^{-1}\{\lambda C_n^o(\lambda)+\lambda\Phi(\lambda)C_n(\lambda)\}\ . \tag{7.13}$$

THEOREM 7.2. *If H_n and H_{n-1} are invertible, then σ_n defines a one to one map of $\mathcal{L}_2(k)$ onto $\mathcal{Q}_2(H_n;k)$, which is inverted by $\hat{\sigma}_n$.*

PROOF. If $F\in\mathcal{L}_2(k)$, then $\lambda^k(A_nF+\lambda C_n^o)$ is both analytic and invertible in a neighborhood of zero, and hence, upon multiplying the top and bottom of the right hand side of (7.12) by λ^k, it follows readily from Lemma 7.2 that

$$\Phi=\sigma_n[F]$$

belongs to $\mathcal{Q}_1$. Moreover, since $F_1=\lambda^kF$ is invertible in a neighborhood of zero and

$$\begin{aligned}(A_n^o-\Phi A_n)F_1 &= A_n^oF_1-\Phi(A_nF_1+\lambda^{k+1}C_n-\lambda^{k+1}C_n)\\ &= A_n^oF_1-(A_n^oF_1-\lambda^{k+1}C_n^o)+\lambda^{k+1}\Phi C_n\\ &= \lambda^{k+1}(C_n^o+\Phi C_n)\ ,\end{aligned}$$

it follows readily from (7.8) that $\Phi\in\mathcal{Q}_2(H_n;k)$. The same formula also serves to show that σ_n is one to one, much as in the proof of the preceding theorem, and that $F=\hat{\sigma}_n[\Phi]$.

Finally, if $\Phi\in\mathcal{Q}_2(H_n;k)$, then it is easily checked that $F=\hat{\sigma}_n[\Phi]$ belongs to $\mathcal{L}_2(k)$ and $\sigma_n[F]=\Phi$. ∎

8. INTERPOLATION IN THE PSEUDO CARATHÉODORY CLASS

A $p \times p$ matrix valued function is said to belong to the Nevanlinna class $\mathcal{N}_{p\times p}$ if each of its entries is the ratio of two scalar H^∞ functions. Matrices in this class have nontangential limits a.e. on the boundary. Clearly $\mathcal{N}_{p\times p} \subset \mathcal{M}_{p\times p}$ and it is readily seen that Theorems 7.1 and 7.2 remain valid if $\mathcal{M}_{p\times p}$ is replaced by $\mathcal{N}_{p\times p}$ in the classes considered there. In this section we shall consider interpolants in the even more restrictive class $\hat{\mathcal{C}}_{p\times p}$ of those $\Phi \in \mathcal{N}_{p\times p}$ for which

$$\Phi(\lambda) + \Phi(\lambda)^* \geq 0 \tag{8.1}$$

for a.e. point $\lambda \in \mathbb{T}$. For obvious reasons, $\hat{\mathcal{C}}_{p\times p}$ is often referred to as the Pseudo Carathéodory class. In particular we shall characterize the classes

$$\mathcal{Q}_3(H_n) = \mathcal{Q}_1(H_n) \cap \hat{\mathcal{C}}_{p\times p}$$

and

$$\mathcal{Q}_4(H_n;k) = \mathcal{Q}_2(H_n;k) \cap \hat{\mathcal{C}}_{p\times p}$$

as linear fractional transformations of the classes

$$\begin{aligned} \mathcal{L}_3(H_n):\quad & G \in \mathcal{L}_1 \cap \mathcal{N}_{p\times p} \ \text{ such that} \\ & \gamma_{00}^{(n)} - G(\lambda)^* \gamma_{nn}^{(n)} G(\lambda) \geq 0\ , \quad \text{for a.e. } \lambda \in \mathbb{T}\ , \end{aligned} \tag{8.2}$$

and

$$\begin{aligned} \mathcal{L}_4(H_n;k):\quad & F \in \mathcal{L}_2(k) \cap \mathcal{N}_{p\times p} \ \text{ such that} \\ & F(\lambda)^* \gamma_{00}^{(n)} F(\lambda) - \gamma_{nn}^{(n)} \geq 0\ , \end{aligned} \tag{8.3}$$

a.e. on $\mathbb{T}$, respectively.

In order to keep the notation simple we shall typically drop the dependence of the spaces $\mathcal{L}_j$ and $\mathcal{Q}_j$ on H_n. Also, from now on we shall say that $G_1(\lambda) = G_2(\lambda)$ at essentially every point in $\mathbb{D}$ if the indicated (matrix) functions agree at every point $\lambda \in \mathbb{D}$ except for an at most countable set of isolated points.

LEMMA 8.1. *If H_n and H_{n-1} are invertible and $G \in \mathcal{L}_1 \cap \mathcal{N}_{p\times p}$, then $\Phi = \tau_n[G]$ belongs to $\mathcal{Q}_1 \cap \mathcal{N}_{p\times p}$ and*

$$\{\Phi(\lambda)C_n(\lambda) + C_n^o(\lambda)\}\lambda G(\lambda) = A_n^o(\lambda) - \Phi(\lambda)A_n(\lambda) \tag{8.4}$$

at essentially every point $\lambda \in \mathbb{D}$. Moreover,

$$(\Phi C_n + C_n^o)(\{\gamma_{nn}^{(n)}\}^{-1} - G\{\gamma_{00}^{(n)}\}^{-1}G^*)(\Phi C_n + C_n^o)^* = 2(\Phi + \Phi^*) \tag{8.5}$$

a.e. on $\mathbb{T}$.

PROOF. Clearly $\Phi \in \mathcal{N}_{p\times p}$ and

$$\begin{aligned} (\Phi C_n + C_n^o)\lambda G &= \Phi(\lambda C_n G + A_n - A_n) + \lambda C_n^o G \\ &= A_n^o - \lambda C_n^o G - \Phi A_n + \lambda C_n^o G \end{aligned}$$

at essentially every point $\lambda \in \mathbb{D}$. This serves to establish (8.4).

Next, it follows readily from (8.4) that

$$\begin{aligned}&\lambda(\Phi C_n + C_n^o)(\{\gamma_{nn}^{(n)}\}^{-1} - G\{\gamma_{00}^{(n)}\}^{-1}G^*)(\Phi C_n + C_n^o)^*\lambda^* \\ &= \lambda[\Phi \quad I_p]\Theta_n\Delta_n^{-1}\Theta_n^*[\Phi \quad I_p]^*\lambda^* \ .\end{aligned}$$

But this tends radially to a limit a.e. on $\mathbb{T}$ which, by (6.4), is equal to

$$[\Phi \quad I_p]J_1[\Phi \quad I_p]^* = 2(\Phi + \Phi^*) \ .$$

∎

LEMMA 8.2. *If H_n and H_{n-1} are invertible and $\Phi \in \mathcal{Q}_1 \cap \mathcal{N}_{p\times p}$, then $G = \hat{\tau}_n[\Phi]$ belongs to $\mathcal{L}_1 \cap \mathcal{N}_{p\times p}$,*

$$A_n^o - \lambda C_n^o G = \Phi(A_n + \lambda C_n G) \tag{8.6}$$

in essentially all of $\mathbb{D}$ and

$$(A_n + \lambda C_n G)^*(\Phi + \Phi^*)(A_n + \lambda C_n G) = 2\{\gamma_{00}^{(n)} - G^*\gamma_{nn}^{(n)}G\} \tag{8.7}$$

a.e. on $\mathbb{T}$.

PROOF. The fact that $G \in \mathcal{L}_1$ is established in the proof of Theorem 7.1. Now, at essentially all points in $\mathbb{D}$,

$$\begin{aligned}A_n^o - \lambda C_n^o G &= A_n^o - (\lambda C_n^o + \lambda\Phi C_n - \lambda\Phi C_n)G \\ &= A_n^o - (A_n^o - \Phi A_n) + \lambda\Phi C_n G \ ,\end{aligned}$$

which clearly implies (8.6).

Finally, (8.6) implies that

$$(A_n + \lambda C_n G)^*(\Phi + \Phi^*)(A_n + \lambda C_n G) = -[G^* \quad I_p]\Theta_n^* J_1 \Theta_n \begin{bmatrix} G \\ I_p \end{bmatrix}$$

in essentially all of $\mathbb{D}$. But this tends radially to

$$-[G^* \quad I_p]2\Delta_n \begin{bmatrix} G \\ I_p \end{bmatrix}$$

a.e. on $\mathbb{T}$ and hence yields (8.7). ∎

LEMMA 8.3. *If X and Y are invertible Hermitian matrices such that $\mu_\pm(X) = \mu_\pm(Y)$, then the matrices $X - Z^*YZ$ and $Y^{-1} - ZX^{-1}Z^*$ have the same inertia.*

PROOF. By Schur complements it is readily seen that the three matrices

$$\begin{bmatrix} X & Z^* \\ Z & Y^{-1} \end{bmatrix}, \quad \begin{bmatrix} X & 0 \\ 0 & Y^{-1} - ZX^{-1}Z^* \end{bmatrix} \quad \text{and} \quad \begin{bmatrix} X - Z^*YZ & 0 \\ 0 & Y^{-1} \end{bmatrix}$$

are congruent to each other. Thus, by Sylvester's law of inertia (see e.g. p.188 of [LT]), they all have the same inertia. Therefore,

$$\mathcal{I}n(X) + \mathcal{I}n(Y^{-1} - ZX^{-1}Z^*) = \mathcal{I}n(X - Z^*YZ) + \mathcal{I}n(Y^{-1}) .$$

But this yields the desired conclusion since

$$\mathcal{I}n(X) = In(Y) = \mathcal{I}n(Y^{-1}) . \qquad \blacksquare$$

It is perhaps of interest to observe that if $X = Y = J$, a signature matrix, then Lemma 8.3 implies that

$$\mathcal{I}n(J - GJG^*) = \mathcal{I}n(J - G^*JG)$$

and hence that

$$J \geq GJG^* \quad \textit{if and only if} \quad J \geq G^*JG , \tag{8.8}$$

a fact which is often proved by more laborious methods.

THEOREM 8.1. *If H_n and H_{n-1} are invertible, then τ_n defines a one to one mapping of $\mathcal{L}_3$ onto $\mathcal{Q}_3$ which is inverted by $\hat{\tau}_n$.*

PROOF. Since $\mathcal{L}_3 \subset \mathcal{L}_1$, it follows readily with the help of Theorem 7.1 that τ_n defines a one to one mapping of $\mathcal{L}_3$ into $\mathcal{Q}_1 \cap \mathcal{N}_{p\times p}$. But now if $G \in \mathcal{L}_3$, then, by Lemmas 4.2 and 8.3,

$$\{\gamma_{nn}^{(n)}\}^{-1} - G\{\gamma_{00}^{(n)}\}^{-1}G^* \geq 0 \tag{8.9}$$

a.e. on $\mathbb{T}$ and hence it follows from (8.5) that $\Phi = \tau_n[G]$ belongs to $\mathcal{Q}_3$. Moreover, $G = \hat{\tau}_n[\Phi]$ by Theorem 7.1.

Next, take any $\Phi \in \mathcal{Q}_3$. Then, since $\mathcal{Q}_3 \subset \mathcal{Q}_1$, $\Phi = \tau_n[G]$ for some $G \in \mathcal{L}_1$ by Theorem 7.1. Clearly $G \in \mathcal{N}_{p\times p}$ and, by another application of Theorem 7.1, $G = \hat{\tau}_n[\Phi]$. Thus (8.7) is applicable and clearly shows that G also satisfies (8.2). $\blacksquare$

LEMMA 8.4. *If H_n and H_{n-1} are invertible and $F \in \mathcal{L}_2(k) \cap \mathcal{N}_{p\times p}$, then $\Phi = \sigma_n[F]$ belongs to $\mathcal{Q}_2(H_n; k) \cap \mathcal{N}_{p\times p}$,*

$$(A_n^o - \Phi A_n)F = \lambda(C_n^o + \Phi C_n) \tag{8.10}$$

in essentially all of $\mathbb{D}$ and

$$(A_n^o - \Phi A_n)(F\{\gamma_{nn}^{(n)}\}^{-1}F^* - \{\gamma_{00}^{(n)}\}^{-1})(A_n^o - \Phi A_n)^* = 2(\Phi + \Phi^*) \tag{8.11}$$

a.e. on $\mathbb{T}$.

PROOF. It follows easily from Theorem 7.2 that Φ belongs to the indicated space. Identity (8.10) is established in the proof of Theorem 7.2. Thanks to it the left hand side of (8.11) can be reexpressed as

$$[\Phi \;\; I_p]\Theta_n\Delta_n^{-1}\Theta_n^*[\Phi \;\; I_p]^*$$

at essentially every point in $\mathbb{D}$. (8.11) emerges upon taking radial limits since the last expression tends radially to

$$2[\Phi \quad I_p]J_1[\Phi \quad I_p]^* = 2(\Phi + \Phi^*)$$

a.e. on $\mathbb{T}$; see (6.4). ∎

LEMMA 8.5. *If H_n and H_{n-1} are invertible and $\Phi \in \mathcal{Q}_2(k) \cap \mathcal{N}_{p\times p}$, then $F = \hat{\sigma}_n[\Phi]$ belongs to $\mathcal{L}_2(k) \cap \mathcal{N}_{p\times p}$,*

$$A_n^o F - \lambda C_n^o = \Phi(A_n F + \lambda C_n) \tag{8.12}$$

in essentially all of $\mathbb{D}$ and

$$(\lambda C_n + A_n F)^*(\Phi + \Phi^*)(\lambda C_n + A_n F) = 2(F^* \gamma_{00}^{(n)} F - \gamma_{nn}^{(n)}) \tag{8.13}$$

a.e. on $\mathbb{T}$.

PROOF. It follows readily from Theorem 7.2 that F belongs to $\mathcal{L}_2(k) \cap \mathcal{N}_{p\times p}$. Moreover, the formula

$$\begin{aligned} A_n^o F - \lambda C_n^o &= (A_n^o - \Phi A_n + \Phi A_n)F - \lambda C_n^o \\ &= \lambda(C_n^o + \Phi C_n) + \Phi A_n F - \lambda C_n^o , \end{aligned}$$

which is valid at essentially every point in $\mathbb{D}$, clearly yields (8.12).

Finally (8.12) serves to identify the left hand side of (8.13) as

$$-[I_p \quad F^*]\Theta_n^* J_1 \Theta_n [I_p \quad F^*]^*$$

essentially everywhere in $\mathbb{D}$ and (8.13) itself emerges upon passing radially to the boundary and invoking (6.2). ∎

THEOREM 8.2. *If H_n and H_{n-1} are invertible, then σ_n defines a one to one map of $\mathcal{L}_4(k)$ onto $\mathcal{Q}_4(k)$ which is inverted by $\hat{\sigma}_n$.*

PROOF. Since $\mathcal{L}_4(k) \subset \mathcal{L}_2(k)$, σ_n defines a one to one map of $\mathcal{L}_4(k)$ into $\mathcal{Q}_2(k)$ which is inverted by $\hat{\sigma}_n$. By Lemmas 4.2 and 8.3, and formula (8.11) it further follows that $\Phi = \sigma_n[F]$ belongs to $\mathcal{Q}_4(k)$ when $F \in \mathcal{L}_4(k)$. On the other hand, if $\Phi \in \mathcal{Q}_4(k)$, then, by Lemma 8.5, $F = \hat{\sigma}_n[\Phi]$ belongs to $\mathcal{L}_2(k) \cap \mathcal{N}_{p\times p}$ and satisfies (8.3), thanks to (8.13). ∎

9. INTERPOLATION IN THE WIENER ALGEBRA

In this section we shall add the constraint that the interpolant belongs to the Wiener algebra $\mathcal{W}_{p\times p}$ of $p \times p$ matrix valued functions

$$f(\lambda) = \sum_{j=-\infty}^{\infty} f_j \lambda^j , \qquad \lambda \in \mathbb{T} ,$$

on the unit circle with

$$\sum_{j=-\infty}^{\infty} |f_j| < \infty .$$

The one sided algebras

$$(\mathcal{W}_{p\times p})_+ = \{f \in \mathcal{W}_{p\times p} : \ f_j = 0 \ \text{ for } \ j < 0\} ,$$

and

$$(\mathcal{W}_{p\times p})_- = \{f \in \mathcal{W}_{p\times p} : \ f_j = 0 \ \text{ for } \ j > 0\} ,$$

will also intervene in the sequel. We begin with the classes

$$\begin{aligned}\mathcal{L}_5(H_n) =& \{G \in \mathcal{L}_1 \cap \mathcal{N}_{p\times p} \cap \mathcal{W}_{p\times p} : \ \gamma_{00}^{(n)} - G(\lambda)^*\gamma_{nn}^{(n)}G(\lambda) \\ & \text{is invertible for every point } \ \lambda \in \mathbb{T}\} \\ \mathcal{Q}_5(H_n) =& \{\Phi \in \mathcal{Q}_1(H_n) \cap \mathcal{N}_{p\times p} \cap \mathcal{W}_{p\times p} : \ \Phi(\lambda) + \Phi(\lambda)^* \ \text{ and} \\ & \Phi C_n + C_n^o \ \text{ are invertible for every point } \ \lambda \in \mathbb{T}\} .\end{aligned}$$

THEOREM 9.1. *If H_n and H_{n-1} are invertible, then τ_n defines a one to one mapping of $\mathcal{L}_5$ onto $\mathcal{Q}_5$ which is inverted by $\hat{\tau}_n$.*

PROOF. The proof is divided into steps.

STEP 1. *$A_n \pm \lambda C_n G$ and $A_n^o \pm \lambda C_n^o G$ are invertible in $\mathcal{W}_{p\times p}$ for every $G \in \mathcal{L}_5$.*

PROOF OF STEP 1. By a well known theorem of Wiener it suffices to prove that the null space of each of the indicated matrices is equal to zero at every point $\lambda \in \mathbb{T}$. Suppose to this end that there exists a vector $\xi \in \mathbb{C}^p$ and a point $\lambda \in \mathbb{T}$ such that

$$\xi^* A_n(\lambda) = \pm\xi^*\lambda C_n(\lambda)G(\lambda) .$$

Then, by (4.19),

$$\xi^* C_n(\lambda)(G(\lambda)\{\gamma_{00}^{(n)}\}^{-1}G(\lambda)^* - \{\gamma_{nn}^{(n)}\}^{-1})C_n(\lambda)^*\xi = 0 .$$

Therefore, $\xi = 0$ since both the middle term and $C_n(\lambda)$ are invertible on $\mathbb{T}$ thanks to Lemma 8.3, the prevailing assumption on G and Corollary 2 to Theorem 6.1. This proves the first pair of assertions. The remaining two may be established in much the same way, using (5.27) in place of (4.19).

STEP 2. *$\Phi = \tau_n[G]$ belongs to $\mathcal{W}_{p\times p}$ and $\Phi(\lambda) + \Phi(\lambda)^*$ is invertible for every point $\lambda \in \mathbb{T}$.*

PROOF OF STEP 2. Step 1 guarantees that $\Phi \in \mathcal{W}_{p\times p}$. Moreover, since

$$[G^* \ \ I_p]\Theta_n^* J_1 \Theta_n \begin{bmatrix} G \\ I_p \end{bmatrix} = 2[G^* \ \ I_p]\Delta_n \begin{bmatrix} G \\ I_p \end{bmatrix}$$

on $\mathbb{T}$ by (6.2), it follows readily upon carrying out the indicated matrix multiplications that (8.7) holds at every point $\lambda \in \mathbb{T}$ and hence that $\Phi + \Phi^*$ is invertible on $\mathbb{T}$.

STEP 3. *If* $\Phi = \tau_n[G]$ *with* $G \in \mathcal{L}_5$, *then* $\Phi C_n + C_n^o$ *is invertible on* $\mathbb{T}$.

PROOF OF STEP 3. By Lemma 8.1, formula (8.5) holds for every point $\lambda \in \mathbb{T}$ (since all the terms in the formula belong to $\mathcal{W}_{p\times p}$ and are therefore continuous on $\mathbb{T}$). But then, since the central term involving G is invertible by assumption (see Lemma 8.3), it follows that at each point $\lambda \in \mathbb{T}$, $\Phi C_n + C_n^o$ is invertible if and only if $\Phi + \Phi^*$ is.

STEP 4. τ_n *defines a one to one mapping of* $\mathcal{L}_5$ *into* $\mathcal{Q}_5$ *which is inverted by* $\hat{\tau}_n$.

PROOF OF STEP 4. By Theorem 7.1, τ_n defines a one to one mapping of $\mathcal{L}_5$ into $\mathcal{Q}_1 \cap \mathcal{N}_{p\times p}$ which is inverted by $\hat{\tau}_n$. The rest is immediate from Steps 2 and 3.

STEP 5. *If* $\Phi \in \mathcal{Q}_5$, *then* $G = \hat{\tau}_n[\Phi]$ *belongs to* $\mathcal{L}_5$.

PROOF OF STEP 5. $\hat{\tau}_n[\Phi]$ belongs to $\mathcal{L}_1$ by Theorem 7.1. It also clearly belongs to $\mathcal{N}_{p\times p}$ as well as to $\mathcal{W}_{p\times p}$, since $\Phi C_n + C_n^o$ is invertible on $\mathbb{T}$. ∎

We turn next to the spaces

$$\begin{aligned}\mathcal{L}_7(H_n) = &\{G \in \mathcal{L}_5 : \ \gamma_{00}^{(n)} - G(\lambda)^* \gamma_{nn}^{(n)} G(\lambda) \\ &\text{is positive definite for every point } \ \lambda \in \mathbb{T}\}\end{aligned}$$

and

$$\begin{aligned}\mathcal{Q}_7(H_n) = &\{\Phi \in \mathcal{Q}_1 \cap \mathcal{N}_{p\times p} \cap \mathcal{W}_{p\times p} : \ \Phi(\lambda) + \Phi(\lambda)^* \\ &\text{is positive definite for every point } \ \lambda \in \mathbb{T}\} \ .\end{aligned}$$

THEOREM 9.2. *If* H_n *and* H_{n-1} *are invertible and* $\gamma_{00}^{(n)}$ *is positive definite, then* τ_n *defines a one to one mapping of* $\mathcal{L}_7$ *onto* $\mathcal{Q}_7$ *which is inverted by* $\hat{\tau}_n$.

PROOF. Since $\mathcal{L}_7 \subset \mathcal{L}_5$, τ_n clearly defines a one to one map of $\mathcal{L}_7$ into $\mathcal{Q}_5$ which is inverted by $\hat{\tau}_n$. Moreover, since $\Phi C_n + C_n^o$ is invertible for every point $\lambda \in \mathbb{T}$, it is immediate from (8.5) that $\Phi + \Phi^* > 0$ on $\mathbb{T}$ and hence that $\Phi \in \mathcal{Q}_7$.

The next step is to show that $\mathcal{Q}_7 \subset \mathcal{Q}_5$. Suppose to this end that

$$\{\Phi(\lambda) C_n(\lambda) + C_n^o(\lambda)\}\xi = 0$$

for some point $\lambda \in \mathbb{T}$ and some vector $\xi \in \mathbb{C}^p$. Then

$$\xi^*\{C_n(\lambda)^*\Phi(\lambda)C_n(\lambda) + C_n(\lambda)^* C_n^o(\lambda)\}\xi = 0 \ .$$

Therefore, upon taking real parts and invoking the formula

$$C_n^* C_n^o + C_n^{o*} C_n \doteq 2\gamma_{nn}^{(n)}$$

which is valid on $\mathbb{T}$ by (6.2), it follows that

$$\xi^*\{C_n(\lambda)^*[\Phi(\lambda) + \Phi(\lambda)^*]C_n(\lambda) + 2\gamma_{nn}^{(n)}\}\xi = 0 \ .$$

But this implies that $\xi = 0$ since, under the present hypotheses, the term inside the curly brackets is positive definite. Thus $\Phi C_n + C_n^o$ is invertible on $\mathbb{T}$ and hence, by Theorem 9.1 (or formula (7.11)), $G = \hat{\tau}_n[\Phi]$ belongs to $\mathcal{L}_5$.

Finally, the positivity which is needed to bring G into $\mathcal{L}_7$, follows from (8.5). ▮

Next we consider the classes

$$\begin{aligned}\mathcal{L}_6(k) =& \{F \in \mathcal{L}_2(k) \cap \mathcal{N}_{p\times p} \cap \mathcal{W}_{p\times p} : \ \gamma_{nn}^{(n)} - F(\lambda)^* \gamma_{00}^{(n)} F(\lambda) \\ & \text{is invertible for every point } \ \lambda \in \mathbb{T}\}\end{aligned}$$

and

$$\begin{aligned}\mathcal{Q}_6(k) =& \{\Phi \in \mathcal{Q}_2(k) \cap \mathcal{N}_{p\times p} \cap \mathcal{W}_{p\times p} : \ \Phi(\lambda) + \Phi(\lambda)^* \text{ and } A_n^o(\lambda) - \Phi(\lambda) A_n(\lambda) \\ & \text{are invertible for every point } \ \lambda \in \mathbb{T}\} \ .\end{aligned}$$

THEOREM 9.3. *If H_n and H_{n-1} are invertible, then σ_n defines a one to one mapping of $\mathcal{L}_6(k)$ onto $\mathcal{Q}_6(k)$ which is inverted by $\hat{\sigma}_n$.*

PROOF. The proof is divided into steps.

STEP 1. *$A_n F \pm \lambda C_n$ and $A_n^o F \pm \lambda C_n^o$ are invertible in $\mathcal{W}_{p\times p}$ for every $F \in \mathcal{L}_6(k)$.*

PROOF OF STEP 1. If

$$\xi^* \{A_n(\lambda) F(\lambda) + \lambda C_n(\lambda)\} = 0$$

for some point $\lambda \in \mathbb{T}$ and some vector $\xi \in \mathbb{C}^p$, then, at this point,

$$\begin{aligned}\xi^* A_n F \{\gamma_{nn}^{(n)}\}^{-1} F^* A_n^* \xi &= \xi^* C_n \{\gamma_{nn}^{(n)}\}^{-1} C_n^* \xi \\ &= \xi^* A_n \{\gamma_{00}^{(n)}\}^{-1} A_n^* \xi \ .\end{aligned}$$

Therefore,

$$\xi^* A_n (F \{\gamma_{nn}^{(n)}\}^{-1} F^* - \{\gamma_{00}^{(n)}\}^{-1}) A_n^* \xi = 0 \ .$$

Thus, since the central term is invertible on $\mathbb{T}$ by Lemma 8.3 and A_n is invertible on $\mathbb{T}$ by Corollary 2 to Theorem 6.1, it follows that $\xi = 0$. This proves the invertibility of $A_n F + \lambda C_n$ on $\mathbb{T}$. The invertibility on $\mathbb{T}$ of the remaining three matrices is established in much the same way.

STEP 2. *If $F \in \mathcal{L}_6(k)$ and $\Phi = \sigma_n[F]$, then $\Phi \in \mathcal{W}_{p\times p}$ and $\Phi + \Phi^*$ is invertible on $\mathbb{T}$.*

PROOF OF STEP 2. Step 1 guarantees that $\Phi \in \mathcal{W}_{p\times p}$. Moreover, since $\mathcal{L}_6(k) \subset \mathcal{L}_2(k)$, $\Phi \in \mathcal{Q}_2(k)$ and hence the rest follows from (8.13).

STEP 3. *σ_n defines a one to one mapping of $\mathcal{L}_6(k)$ into $\mathcal{Q}_6(k)$ which is inverted by $\hat{\sigma}_n$.*

PROOF OF STEP 3. Since $\mathcal{L}_6(k) \subset \mathcal{L}_2(k)$, σ_n clearly defines a one to one mapping of $\mathcal{L}_6(k)$ into $\mathcal{Q}_2(k) \cap \mathcal{N}_{p\times p}$ which is inverted by $\hat{\sigma}_n$. The rest is immediate from Step 2 and formula (8.11).

STEP 4. *If $\Phi \in \mathcal{Q}_6(k)$, then $F = \hat{\sigma}_n[\Phi]$ belongs to $\mathcal{L}_6(k)$.*

PROOF OF STEP 4. Since $\mathcal{Q}_6(k) \subset \mathcal{Q}_2(k)$, $F \in \mathcal{L}_2(k)$ by Theorem 7.2. Next, since $A_n^o - \Phi A_n$ is invertible on $\mathbb{T}$ by assumption, $F \in \mathcal{W}_{p\times p}$ as well as to $\mathcal{N}_{p\times p}$. Finally, since $\Phi = \sigma_n[F]$, it follows from (8.11) and Lemma 8.3 that $\gamma_{nn}^{(n)} - F^*\gamma_{00}^{(n)}F$ is invertible on $\mathbb{T}$, as needed to complete the proof of the step and the theorem. ∎

Now let

$$\mathcal{L}_8(k) = \{F \in \mathcal{L}_6(k):\ \gamma_{nn}^{(n)} - F(\lambda)^*\gamma_{00}^{(n)}F(\lambda) \text{ is negative definite for every point } \lambda \in \mathbb{T}\}$$

and

$$\mathcal{Q}_8(k) = \{\Phi \in \mathcal{Q}_2(k) \cap \mathcal{N}_{p\times p} \cap \mathcal{W}_{p\times p}:\ \Phi(\lambda) + \Phi(\lambda)^* \text{ is positive definite for every point } \lambda \in \mathbb{T}\}\ .$$

THEOREM 9.4. *If H_n and H_{n-1} are invertible and if $\gamma_{00}^{(n)}$ is negative definite, then σ_n defines a one to one map of $\mathcal{L}_8(k)$ onto $\mathcal{Q}_8(k)$ which is inverted by $\hat{\sigma}_n$.*

PROOF. Since $\mathcal{L}_8(k) \subset \mathcal{L}_6(k)$, Theorem 9.3 guarantees that σ_n defines a one to one map of $\mathcal{L}_8(k)$ into $\mathcal{Q}_6(k)$ which is inverted by $\hat{\sigma}_n$. Moreover, if $\Phi = \sigma_n[F]$ with $F \in \mathcal{L}_8(k)$, it follows from Lemma 8.3 and (8.11) that $\Phi + \Phi^* > 0$ on $\mathbb{T}$.

The next step is to show that under the present hypotheses $\mathcal{Q}_8(k) \subset \mathcal{Q}_6(k)$. To do this we have to show that $A_n^o - \Phi A_n$ is invertible on $\mathbb{T}$ for $\Phi \in \mathcal{Q}_8(k)$. But if

$$\{A_n^o(\lambda) - \Phi(\lambda)A_n(\lambda)\}\xi = 0$$

for some point $\lambda \in \mathbb{T}$ and some vector $\xi \in \mathbb{C}^p$, then, at this point,

$$\xi^*\{A_n^*\Phi A_n - A_n^*A_n^o\}\xi = 0$$

and therefore, upon taking the real part and invoking the identity

$$A_n^*A_n^o + A_n^{o*}A_n = 2\gamma_{00}^{(n)}\ ,$$

which is valid on $\mathbb{T}$ by (6.2), it follows readily that

$$\xi^*\{A_n^*(\Phi + \Phi^*)A_n - 2\gamma_{00}^{(n)}\}\xi = 0\ .$$

But this forces $\xi = 0$, since the term in curly brackets is positive definite. Thus $A_n^o - \Phi A_n$ is invertible on $\mathbb{T}$ and so $F = \hat{\sigma}_n[\Phi]$ belongs to $\mathcal{L}_6(k)$ for $\Phi \in \mathcal{Q}_8(k)$ and $\Phi = \sigma_n[F]$. The rest is plain from (8.11) and Lemma 8.3. ∎

10. THE COVARIANCE EXTENSION PROBLEM

Let $\mathcal{P}(H_n)$ denote the set of $p \times p$ matrix valued functions

$$f(\lambda) = \sum_{-\infty}^{\infty} f_j\lambda^j$$

in the Wiener algebra $\mathcal{W}_{p\times p}$ such that

(1) $f_j = h_j,\ j = -n, \ldots, n$, and

(2) $f(\lambda)$ is positive definite for every point $\lambda \in \mathbb{T}$.

THEOREM 10.1. *$\mathcal{P}(H_n)$ is nonempty if and only if H_n is positive definite.*

PROOF. Suppose first that $\mathcal{P}$ is nonempty. Then for any $f \in \mathcal{P}$ and any choice of vectors $\xi_0, \ldots, \xi_n$ in $\mathbb{C}^p$,

$$\begin{aligned}\sum_{j,k=0}^{n} \xi_j^* h_{k-j} \xi_j &= \sum_{j,k=0}^{n} \xi_j^* f_{k-j} \xi_j \\ &= \frac{1}{2\pi} \int_0^{2\pi} \left\{ \sum_{j=0}^{n} \xi_j e^{ij\theta} \right\}^* f(e^{i\theta}) \left\{ \sum_{k=0}^{n} \xi_k e^{ik\theta} \right\} d\theta \\ &> 0\end{aligned}$$

unless all the vectors ξ_j, $j = 0, \ldots, n$, are identically zero. This proves the necessity of the condition $H_n > 0$ in order for $\mathcal{P}$ to be nonempty.

Suppose next that $H_n > 0$. Then H_n and H_{n-1} are invertible and $\gamma_{00}^{(n)} > 0$. Therefore $G(\lambda) \equiv 0$ belongs to $\mathcal{L}_5$ and, by Theorem 9.1, $\Phi = \tau_n[0] = A_n^o A_n^{-1}$ belongs to $\mathcal{Q}_5(H_n)$. In fact, since $A_n(\lambda)$ is invertible for every point $\lambda \in \bar{\mathbb{D}}$ by Lemma 4.4, it follows from a well-known theorem of Wiener that $\Phi \in (\mathcal{W}_{p\times p})_+$ and hence, by a simple calculation, that

$$f(\lambda) = \frac{\Phi(\lambda) + \Phi(\lambda)^*}{2}, \qquad \lambda \in \mathbb{T}$$

belongs to $\mathcal{P}(H_n)$; (8.7) guarantees the requisite positivity on $\mathbb{T}$. ∎

It is now expedient to introduce the classes

$$\mathcal{L}_9 = \begin{array}{l}\{G \in (\mathcal{W}_{p\times p})_+ :\ \gamma_{00}^{(n)} - G(\lambda)^* \gamma_{nn}^{(n)} G(\lambda) \ \text{ is} \\ \text{positive definite for every point } \lambda \in \mathbb{T}\}\end{array}$$

and

$$\mathcal{Q}_9 = \begin{array}{l}\{\Phi \in \mathcal{Q}_1 \cap (\mathcal{W}_{p\times p})_+ :\ \Phi(\lambda) + \Phi(\lambda)^* \ \text{ is} \\ \text{positive definite for every point } \lambda \in \mathbb{T}\}\end{array}$$

THEOREM 10.2. *The class $\mathcal{Q}_9(H_n)$ is nonempty if and only if H_n is positive definite.*

PROOF. If $H_n > 0$, then, by Theorem 10.1, $\mathcal{P}$ is nonempty. Let $f = \sum_{j=-\infty}^{\infty} f_j \lambda^j$ belong to $\mathcal{P}$. Then clearly $f_j = f_{-j}^*$ and

$$\Phi = f_0 + 2 \sum_{j=1}^{\infty} f_j \lambda^j$$

belongs to $\mathcal{Q}_9$. Thus $\mathcal{Q}_9$ is nonempty.

Conversely, if $\mathcal{Q}_9$ is nonempty, then

$$f = (\Phi + \Phi^*)/2 \text{ belongs to } \mathcal{P}(H_n)$$

and therefore, by another application of Theorem 10.1, it follows that H_n is positive definite. ∎

THEOREM 10.3. *If H_n is positive definite, then τ_n defines a one to one map of $\mathcal{L}_9(H_n)$ onto $\mathcal{Q}_9(H_n)$.*

PROOF. Since $\mathcal{L}_9 \subset \mathcal{L}_7$, it follows from Theorem 9.2 that τ_n is a one to one map of $\mathcal{L}_9$ into $\mathcal{Q}_7$. The rest of the proof proceeds in steps.

STEP 1. *If $G \in \mathcal{L}_9$, then $A_n(\lambda) + \lambda C_n(\lambda)G(\lambda)$ is invertible for every point $\lambda \in \bar{\mathbb{D}}$.*

PROOF OF STEP 1. Since $A_n(0)$ is invertible and, by (8.7), $A_n + \lambda C_n G$ is invertible on $\mathbb{T}$, it remains only to consider nonzero λ in $\mathbb{D}$. Suppose, that for such a point,

$$\xi^* A_n + \xi^* \lambda C_n G = 0$$

for some $\xi \in \mathbb{C}^p$. Then

$$\begin{aligned} &\xi^* \lambda C_n (G\{\gamma_{00}^{(n)}\}^{-1} G^* - \{\gamma_{nn}^{(n)}\}^{-1}) C_n^* \lambda^* \xi \\ &= \xi^* (A_n \{\gamma_{00}^{(n)}\}^{-1} A_n^* - \lambda C_n \{\gamma_{nn}^{(n)}\}^{-1} C_n^* \lambda^*) \xi \\ &> 0 \end{aligned}$$

by (4.19), unless $\xi = 0$. But, since $G \in \mathcal{L}_9$, $\xi = 0$ is the only viable possibility. Therefore $A_n + \lambda C_n G$ is invertible in all of $\bar{\mathbb{D}}$.

STEP 2. *τ_n is a one to one mapping of $\mathcal{L}_9$ into $\mathcal{Q}_9$.*

PROOF OF STEP 2. By Step 1, $(A_n + \lambda C_n G)^{-1}$ belongs to $(\mathcal{W}_{p\times p})_+$. Thus $\tau_n[G] \in (\mathcal{W}_{p\times p})_+$ for every $G \in \mathcal{L}_9$. The rest is plain from the remarks preceding the statement of Step 1.

STEP 3. *If $X = [\Phi \quad I_p]$ with $\Phi \in \mathcal{Q}_9$ and if U is defined as in (2.6), then the matrix*

$$X(\omega)U(\omega)J_1U(\omega)^*X(\omega)^*$$

is positive definite for every point $\omega \in \bar{\mathbb{D}}$ except for $\omega = 0$.

PROOF OF STEP 3. Since Φ is an interpolant the functions f_j which are defined by (2.5) can be expressed in the form

$$f_j = \underline{\underline{p}} J_1 X^* u_j \ , \quad \text{with } u_j(\lambda) = \lambda^j I_p \ ,$$

for $j = 0, \ldots, n$. Therefore, by Cauchy's formula,

$$\xi^* X(\omega) f_i(\omega) \eta = \langle \Omega u_i \eta, \xi / \rho_\omega \rangle$$

where

$$\Omega = \Phi + \Phi^* .$$

Moreover, since

$$< J_1 f_j, f_i > = < \Omega u_j, u_i >$$

it follows from (2.4) and the preceding calculation that

$$\begin{aligned} &\xi^* X(\omega) \frac{J_1 - U(\omega) J_1 U(\omega)^*}{\rho_\omega(\omega)} X(\omega)^* \xi \\ &= \sum_{i,j=0}^{n} < \Omega u_i, \xi/\rho_\omega > (\mathbb{P}^{-1})_{ij} < \xi/\rho_\omega, \Omega u_j > \\ &= < \Omega P \xi/\rho_\omega, P\xi/\rho_\omega > \end{aligned}$$

where P denotes the orthogonal projection of the span onto the columns of the u_i, $i = 0, \ldots, n$, in the Hilbert space of $p \times 1$ vector valued functions with inner product $< \Omega \cdot, \cdot >$. Clearly

$$\begin{aligned} < \Omega P \xi/\rho_\omega, P\xi/\rho_\omega > &\leq < \Omega \xi/\rho_\omega, \xi/\rho_\omega > \\ &= \xi^* \left\{ \frac{X(\omega) J_1 X(\omega)^*}{\rho_\omega(\omega)} \right\} \xi . \end{aligned}$$

Thus, upon combining estimates, it is readily seen that

$$X(\omega) U(\omega) J_1 U(\omega)^* X(\omega)^* \geq 0$$

for every point $\omega \in \bar{\mathbb{D}}$. But now as U is invertible for every nonzero point $\omega \in \bar{\mathbb{D}}$ by (2.7) and X is of rank p everywhere on $\bar{\mathbb{D}}$, the inequality must be strict for every nonzero point $\omega \in \bar{\mathbb{D}}$.

STEP 4. *If $\Phi \in \mathcal{Q}_9$, then the matrix valued function*

$$\{\Phi(\lambda) C_n(\lambda) + C_n^o(\lambda)\}/\lambda^n$$

is invertible for every point $\lambda \in \bar{\mathbb{D}}$.

PROOF OF STEP 4. By (6.4) and (6.5),

$$\begin{aligned} 2X(\omega) U(\omega) J_1 U(\omega)^* X(\omega)^* &= X(\omega) U(\omega) \Theta_n(\alpha) \Delta_n^{-1} \Theta_n(\alpha)^* U(\omega)^* X(\omega)^* \\ &= X(\omega) \Theta_n(\omega) \Delta_n^{-1} \Theta_n(\omega)^* X(\omega)^* \end{aligned}$$

for every $\omega \in \bar{\mathbb{D}}$. Therefore, since this matrix is positive definite by Step 3 if $\omega \neq 0$, it is readily checked by direct computation that

$$\{\Phi(\omega) C_n(\omega) + C_n^o(\omega)\} \{\gamma_{nn}^{(n)}\}^{-1} \{\Phi(\omega) C_n(\omega) + C_n^o(\omega)\} > 0$$

for $\omega \neq 0$. Thus

$$\{\Phi(\lambda) C_n(\lambda) + C_n^o(\lambda)\}/\lambda^n$$

is invertible for every nonzero point $\lambda \in \bar{\mathbb{D}}$. But since Φ is an interpolant, the same is true for $\lambda = 0$ also, thanks to (7.8).

STEP 5. *If $\Phi \in \mathcal{Q}_9$, then $\hat{\tau}_n[\Phi] \in \mathcal{L}_9$.*

PROOF OF STEP 5. Since $\mathcal{Q}_9 \subset \mathcal{Q}_7$, $\hat{\tau}_n[\Phi]$ clearly belongs to $\mathcal{L}_7$. Moreover, in view of Step 4,

$$\hat{\tau}_n[\Phi] = \left\{\frac{\Phi(\lambda)C_n(\lambda) + C_n^o(\lambda)}{\lambda^n}\right\}^{-1}\left\{\frac{A_n^o(\lambda) - \Phi(\lambda)A_n(\lambda)}{\lambda^{n+1}}\right\}$$

also belongs to $(\mathcal{W}_{p\times p})_+$. This completes the proof of the step and the theorem since $\mathcal{L}_9 = \mathcal{L}_7 \cap (\mathcal{W}_{p\times p})_+$. ∎

THEOREM 10.4. *Let H_n be positive definite. Then $f \in \mathcal{P}(H_n)$ if and only if it can be expressed in the form*

$$f = (E_n^*)^{-1}(I_p + \lambda^* S^* V_n^*)^{-1}(I_p - S^*S)(I_p + \lambda V_n S)^{-1}(E_n)^{-1} \tag{10.20}$$

on $\mathbb{T}$, or equivalently if and only if it can be expressed in the form

$$f = F_n^{*-1}(I_p + \lambda S V_n)^{-1}(I_p - SS^*)(I_p + \lambda^* V_n^* S^*)^{-1} F_n^{-1} \tag{10.21}$$

on $\mathbb{T}$, where

$$E_n(\lambda) = \left\{\sum_{i=0}^{n} \lambda^i \gamma_{i0}^{(n)}\right\}\{\gamma_{00}^{(n)}\}^{-1/2} = A_n(\lambda)\{\gamma_{00}^{(n)}\}^{-1/2} \tag{10.22}$$

$$F_n(\lambda) = \left\{\sum_{i=0}^{n} \lambda^i\ \gamma_{in}^{(n)}\right\}\{\gamma_{nn}^{(n)}\}^{-1/2} = C_n(\lambda)\{\gamma_{nn}^{(n)}\}^{-1/2}\ , \tag{10.23}$$

V_n is the Blaschke-Potapov product defined by

$$V_n(\lambda) = E_n(\lambda)^{-1}F_n(\lambda) \tag{10.24}$$

and S is an arbitrary $p\times p$ matrix valued analytic function on $\mathbb{D}$ which belongs to $(\mathcal{W}_{p\times p})_+$ and is strictly contractive on $\bar{\mathbb{D}}$.

PROOF. Since $H_n > 0$ it follows readily from Theorems 10.2 and 10.3 that $f \in \mathcal{P}$ if and only if it can be expressed in the form

$$f = \{\tau_n[G] + \tau_n[G]^*\}/2$$

on $\mathbb{T}$ for some choice of $G \in \mathcal{L}_9$. But now $G \in \mathcal{L}_9$ if and only if

$$S(\lambda) = \{\gamma_{nn}^{(n)}\}^{\frac{1}{2}} G(\lambda)\{\gamma_{00}^{(n)}\}^{-\frac{1}{2}}$$

belongs to $(\mathcal{W}_{p\times p})_+$ and is strictly contractive for every point $\lambda \in \bar{\mathbb{D}}$. Consequently, by (8.7), with $\Phi = \tau_n[G]$, it follows that

$$(E_n + \lambda F_n S)^* f (E_n + \lambda F_n S) = I_p - S^*S$$

on $\mathbb{T}$. This makes the first representation plain since E_n is invertible on $\bar{\mathbb{D}}$ and

$$E_n E_n^* = F_n F_n^*$$

on $\mathbb{T}$ by Lemma 4.4 and (4.19), respectively. The second follows from the first and the identity

$$(I_p + X^*)^{-1}(I_p - X^*X)(I_p + X)^{-1} = (I_p + X)^{-1}(I_p - XX^*)(I_p + X^*)^{-1}$$

which is valid for any strictly contractive $p \times p$ matrix X. Just choose $X = \lambda V_n S$ and calculate away, bearing in mind that λV_n is unitary on $\mathbb{T}$. ∎

Representation formulas of the type exhibited in Theorem 10.4 for the solutions of the covariance extension problem (albeit in a different setting) first appear in Youla [Y]. They are useful in maximum entropy estimates. See e.g., Theorem 11.3 and its Corollary in [D] for the statement. Additional references and discussion are provided in the notes to that chapter.

11. INTERIOR ROOTS

In this section we shall use a part of the preceding analysis to give a new proof of a theorem of Alpay and Gohberg [AG] which expresses the number of roots of $\det\{\lambda C_n(\lambda)\}$ inside $\mathbb{D}$ in terms of $\mu_\pm(H_n)$ when $\gamma_{00}^{(n)}$ is definite. The corresponding result for scalar polynomials was first established by Krein [K2].

We first prepare some lemmas which are of interest in their own right. Therein we shall use the symbol $\boldsymbol{F}$ to denote the Toeplitz operator based on the $p \times p$ matrix polynomial $F(\lambda)$ which is defined by the rule

$$\boldsymbol{F}u = \underline{\underline{p}}F^*u\ , \qquad u \in H_p^2\ .$$

LEMMA 11.1. *If $F(\lambda)$ is a $p \times p$ matrix polynomial which is invertible for every point $\lambda \in \mathbb{T}$, then the number of roots of $\det\{F(\lambda)\}$ inside $\mathbb{D}$ is equal to the dimension of the kernel of $\boldsymbol{F}$.*

PROOF. Under the given assumptions F admits a factorization of the form

$$F(\lambda) = F_-(\lambda)D(\lambda)F_+(\lambda)$$

where $F_\pm$ are invertible in the Wiener subalgebras $(\mathcal{W}_{p\times p})_\pm$ and

$$D(\lambda) = \operatorname{diag}\{\lambda^{\kappa_1}, \lambda^{\kappa_2}, \ldots, \lambda^{\kappa_p}\}$$

with

$$0 \le \kappa_1 \le \kappa_2 \le \ldots \le \kappa_p\ ;$$

see e.g., pages 14-17 of Clancey and Gohberg [CG]. Thus the dimension of the kernel of $\boldsymbol{F}$ is equal to

$$\nu = \kappa_1 + \ldots + \kappa_p$$

and hence upon taken determinants in the factorization formula for F it follows, in a selfevident notion, that

$$\varphi(\lambda) = \varphi_-(\lambda)\lambda^\nu \varphi_+(\lambda) .$$

But this in turn implies by the argument principle that ν is equal to the number of zeros of φ inside $\mathbb{D}$ counting multiplicities. ∎

LEMMA 11.2. *If F is a matrix polynomial, then the kernel of the associated Toeplitz operator $\boldsymbol{F}$ is an R_0 invariant subspace of H_p^2.*

PROOF. Let $u \in \ker \boldsymbol{F}$ and set $\underline{\underline{q}}' = I - \underline{\underline{p}}$. Then clearly

$$\begin{aligned}\underline{\underline{p}}F^* R_0 u &= \underline{\underline{p}}\left\{\frac{1}{\lambda}\underline{\underline{p}}F^* u + \frac{1}{\lambda}\underline{\underline{q}}' F^* u\right\} \\ &= \underline{\underline{p}}\frac{1}{\lambda}\underline{\underline{q}}' F^* u \\ &= 0 .\end{aligned}$$

∎

Lemma 11.2 implies that the kernel of $\boldsymbol{F}$ can be organized into chains of the form

$$g_{ji}(\lambda) = \frac{\lambda^i \xi_{j0}}{\rho_{\alpha_j}(\lambda)^{i+1}} + \frac{\lambda^{i-1}\xi_{j1}}{\rho_{\alpha_j}(\lambda)^i} + \ldots + \frac{\xi_{ji}}{\rho_{\alpha_j}(\lambda)}$$

for $i = 0, \ldots, k_j$, and $j = 1, \ldots, r$, where

$(1 + k_1) + \ldots + (1 + k_r) = \dim \ker \boldsymbol{F}$,

α_j , $j = 1, \ldots, r$ are points in $\mathbb{D}$,

ξ_{ji} , $i = 0, \ldots, k_j$, $j = 1, \ldots, r$ are vectors in $\mathbb{C}^p$

and

$\operatorname{span}\{g_{ji}$, $i = 0, \ldots, k_j$, $j = 1, \ldots, r\} = \ker \boldsymbol{F}$.

LEMMA 11.3. *If X and Y are matrix polynomials such that*

$$X(\lambda)JX(\omega)^* = \rho_\omega(\lambda)Y(\lambda)\Gamma Y(\omega)^* \tag{11.1}$$

and if

$$g(\lambda) = \frac{\lambda^j \xi_0}{\rho_\alpha(\lambda)^{j+1}} + \ldots + \frac{\xi_j}{\rho_\alpha(\lambda)} ,$$

$$h(\lambda) = \frac{\lambda^i \eta_0}{\rho_\beta(\lambda)^{i+1}} + \ldots + \frac{\eta_i}{\rho_\beta(\lambda)} ,$$

where $\xi_0,\ldots,\xi_j;\ \eta_0,\ldots,\eta_i$ *belong to* $\mathbb{C}^p$, α *and* β *are points in* $\mathbb{D}$, *and*

$$u = < Y, g >^* = \frac{Y^{(j)}(\alpha)^*}{j!}\xi_0 + \ldots + Y(\alpha)^*\xi_j$$

$$v = < Y, h >^* = \frac{Y^{(i)}(\beta)^*}{i!}\eta_0 + \ldots + Y(\beta)^*\eta_i \ ,$$

then

$$< XJ\underline{\underline{p}}X^*g, h > = v^*\Gamma u \ .$$

PROOF. Let

$$\varphi_{\alpha,t}(\lambda) = \frac{1}{t!}\left(\frac{\partial}{\partial\alpha^*}\right)^t \frac{1}{\rho_\alpha(\lambda)} = \frac{\lambda^t}{\rho_\alpha(\lambda)^{t+1}} \ , \qquad t = 0, 1, \ldots, \ . \tag{11.2}$$

Then, by Cauchy's formula,

$$\begin{aligned}(\underline{\underline{p}}X^*\varphi_{\alpha,t})(\lambda) &= \frac{1}{t!}\left(\underline{\underline{p}}X^*\left(\frac{\partial}{\partial\alpha^*}\right)^t\varphi_{\alpha,0}\right)(\lambda)\\ &= \frac{1}{t!}\left(\frac{\partial}{\partial\alpha^*}\right)^t(\underline{\underline{p}}X^*\varphi_{\alpha,0})(\lambda)\\ &= \frac{1}{t!}\left(\frac{\partial}{\partial\alpha^*}\right)^t X(\alpha)^*\varphi_{\alpha,0}(\lambda)\end{aligned}$$

and hence

$$\begin{aligned}X(\lambda)J(\underline{\underline{p}}X^*\varphi_{\alpha,t})(\lambda) &= \frac{1}{t!}\left(\frac{\partial}{\partial\alpha^*}\right)^t\frac{X(\lambda)JX(\alpha)^*}{\rho_a(\lambda)}\\ &= \frac{1}{t!}\left(\frac{\partial}{\partial\alpha^*}\right)^t Y(\lambda)\Gamma Y(\alpha)^*\\ &= \frac{Y(\lambda)\Gamma Y^{(t)}(\alpha)^*}{t!} \ .\end{aligned}$$

This identity is the main ingredient in the proof. Thus,

$$\begin{aligned}X(\lambda)J(\underline{\underline{p}}X^*g)(\lambda) &= \sum_{s=0}^{j} Y(\lambda)\Gamma\frac{Y^{(j-s)}(\alpha)^*\xi_s}{(j-s)!}\\ &= Y(\lambda)\Gamma u\end{aligned}$$

and the rest is plain since

$$< Y(\lambda)\Gamma u, h > = v^*\Gamma u \ . \qquad \blacksquare$$

THEOREM 11.1. *If* H_n *and* H_{n-1} *are invertible and* $\gamma_{00}^{(n)}$ *is definite, then* $\det\{\lambda C_n(\lambda)\}$ *has*

$\mu_+(H_n)$ *[resp.* $\mu_-(H_n)$*] roots inside* $\mathbb{D}$

$\mu_-(H_n)$ [resp. $\mu_+(H_n)$] roots inside $\mathbb{E}$

if $\gamma_{00}^{(n)}$ is positive [resp. negative] definite.

PROOF. Suppose first that $\gamma_{00}^{(n)}$ is positive definite. Then, by Lemma 4.2, $\gamma_{nn}^{(n)}$ is also positive definite and therefore the identity (4.19) can be reexpressed in the form (11.1) with $J = J_0$, $\Gamma = \Gamma_n$,

$$X(\lambda) = [A_n(\lambda)\{\gamma_{00}^{(n)}\}^{-\frac{1}{2}} \qquad \lambda C_n(\lambda)\{\gamma_{nn}^{(n)}\}^{-\frac{1}{2}}]$$

and

$$Y(\lambda) = [I_p \quad \lambda I_p \quad \dots \quad \lambda^n I_p] \; .$$

Next, let $F(\lambda) = \lambda C_n(\lambda)$. Then, since F is invertible on $\mathbb{T}$ by Corollary 2 to Theorem 6.1, Lemma 11.1 is applicable. Let ν =dim ker $\boldsymbol{F}$ and let us reindex the chains

$$g_{10}, \dots, g_{1k_1}; \dots ; g_{r0}, \dots, g_{rk_r}$$

which span out the kernel of $\boldsymbol{F}$ by $g_1, \dots, g_\nu$. But now, upon setting

$$u_s = < Y, g_s >^* \; ,$$

it follows from Lemma 11.3 that, for any choice of constants $b_1, \dots, b_\nu$,

$$\begin{aligned} < \Gamma_n \sum_{s=1}^{\nu} b_s u_s, \sum_{t=1}^{n} b_t u_t > &= < X J_0 \underline{\underline{p}} X^* \sum_{s=1}^{\nu} b_s g_s, \sum_{t=1}^{\nu} b_t g_t > \\ &= \left\| \{\gamma_{00}^{(n)}\}^{-\frac{1}{2}} \underline{\underline{p}} A_n^* \sum_{s=1}^{\nu} b_s g_s \right\|^2 \; . \end{aligned}$$

Therefore, the span of the vectors $u_1, \dots, u_\nu$ is a positive subspace of $\mathbb{C}^{(n+1)p}$ with respect to the indefinite inner product induced by Γ_n. This proves that

$$\nu \le \mu_+(\Gamma_n) \; . \tag{11.3}$$

By Lemma 11.1, this exhibits $\mu_+(\Gamma_n)$ as an upper bound on the number of roots of det $F(\lambda)$ inside $\mathbb{D}$.

The next step is to show that $\mu_-(\Gamma_n)$ is an upper bound for the number of roots of det $F(\lambda)$ in $\mathbb{E}$. This is obtained by reexpressing formula (4.19) in terms of the reciprocal polynomials

$$\hat{A}_n(\lambda) = \lambda^n A_n(1/\lambda) \text{ and } \hat{C}_n(\lambda) = \lambda^n C_n(1/\lambda) \tag{11.4}$$

as

$$\hat{C}_n(\lambda)\{\gamma_{nn}^{(n)}\}^{-1}\hat{C}_n(\omega)^* - \lambda \hat{A}_n(\lambda)\{\gamma_{00}^{(n)}\}^{-1}\hat{A}_n(\omega)^*\omega^* = \rho_\omega(\lambda)[\lambda^n I_p \dots I_p]\Gamma_n[\omega^n I_p \dots I_p]^* \; .$$

This is a good thing to do because $\omega \in \mathbb{E}$ is a root of det$\{\lambda C_n(\lambda)\}$ of multiplicity k if and only if $1/\omega$ is a root of det$\{\hat{C}_n(\lambda)\}$ of multiplicity k. Therefore, since $1/\omega$ belongs to $\mathbb{D}$, the preceding argument is applicable and (because of the change in sign) leads readily to the conclusion that if ν' denotes the number of zeros of $\det\{\lambda C_n(\lambda)\}$

$$\begin{aligned}\nu' &= \text{the number of zeros of } \det\{\lambda C_n(\lambda)\} \text{ in } \mathbb{E} \\ &= \text{the number of zeros of } \det\{\hat{C}_n(\lambda)\} \text{in } \mathbb{D}\ ,\end{aligned}$$

is subject to the bound

$$\nu' \leq \mu_-(\Gamma_n)\ . \tag{11.5}$$

Therefore, since $\lambda C_n(\lambda)$ is invertible on $\mathbb{T}$ and its top coefficient $\gamma_{nn}^{(n)}$ is an invertible matrix,

$$(n+1)p = \nu + \nu' \leq \mu_+(\Gamma_n) + \mu_-(\Gamma_n) = (n+1)p\ .$$

Thus equality must prevail in the bounds (11.3) and (11.5). This completes the proof for $\gamma_{00}^{(n)} > 0$. The proof for $\gamma_{00}^{(n)} < 0$ is carried out in much the same way. ∎

THEOREM 11.2 *If H_n and H_{n-1} are invertible and $\gamma_{00}^{(n)}$ is definite, then* $\det\{C_n(\lambda)\}$ *has*

$\mu_+(H_{n-1})$ *[resp. $\mu_-(H_{n-1})$] roots inside* $\mathbb{D}$

$\mu_-(H_{n-1})$ *[resp. $\mu_+(H_{n-1})$] roots inside* $\mathbb{E}$

if $\gamma_{00}^{(n)}$ is positive [resp. negative] definite.

PROOF. This is clearly equivalent to Theorem 11.1, thanks to Lemma 4.2. ∎

Since $\mu_+(H_{n-1}) + \mu_-(H_{n-1}) = np$, Theorem 11.2 corresponds precisely to the statement of the theorem of Alpay and Gohberg in [AG].

12. ISOMETRIES AND NEGATIVE SQUARES

In this section we shall sketch the connection between the number of negative squares of a pair of kernels based on Φ and G for $\Phi = \tau_n[G]$ with $G \in \mathcal{L}_1$. The main tool is Theorem 6.13 of [AD]. We shall be brief, both because the Editor is pressing for the manuscript, and also because a more elaborate study which will cover the present material is planned with D. Alpay.

Let

$$X(\lambda) = [\Phi(\lambda)\ \ I_p]\ ,$$

where $\Phi \in \mathcal{Q}_1(H_n)$, and suppose that the kernel

$$\Lambda_\omega^X(\lambda) = \frac{X(\lambda)J_1X(\omega)^*}{\rho_\omega(\lambda)} = \frac{\Phi(\lambda)+\Phi(\omega)^*}{\rho_\omega(\lambda)} \tag{12.1}$$

has ν negative squares. This means that for every choice of points $\alpha_1, \ldots, \alpha_n$ in the domain of analyticity Ω_X of X in $\mathbb{D}$ and every choice $\xi_1, \ldots, \xi_n$ of vectors in $\mathbb{C}^p$ the $n \times n$ matrix with ij entry equal to

$$\xi_i^* \Lambda_{\alpha_j}^X(\alpha_i)\xi_j\ , \qquad i,j = 1, \ldots, n\ ,$$

has at most ν (strictly) negative eigenvalues and moreover that for some choice of $\alpha_1, \ldots, \alpha_n$ and $\xi_1, \ldots, \xi_n$ it has exactly ν negative eigenvalues. Then there is a uniquely associated reproducing kernel Pontryagin space of $k \times 1$ vector valued meromorphic functions in $\mathbb{D}$ (which are analytic in Ω_X) with reproducing kernel $\Lambda_\omega^X(\lambda)$ given as in (12.1); see [AD]. We shall refer to this space as $\mathcal{B}(X)$. It is R_0 invariant (as follows by identifying $\mathcal{B}(X)$ with a related "$\mathcal{K}(S)$" space and invoking Theorem 6.7 of [AD]).

THEOREM 12.1. *Suppose that H_n and H_{n-1} are invertible, that U and $\mathcal{M}$ are as in Theorem 2.3 and that the kernel $\Lambda_\omega^X(\lambda)$ based on $\Phi \in \mathcal{Q}_1(H_n)$ has ν negative squares. Then $\mathcal{B}(XU)$ sits isometrically inside $\mathcal{B}(X)$ and*

$$\mathcal{B}(X) = \mathcal{B}(XU) \boxplus X\mathcal{M}$$

where the indicated sum is both direct and orthogonal in $\mathcal{B}(X)$.

PROOF. By Theorem 6.13 of [AD] it suffices to show that

(1) $$Xf_j\xi \in \mathcal{B}(X) \quad \text{and}$$

(2) $$\left\| \sum_{j=0}^{n} Xf_j\xi_j \right\|_{\mathcal{B}(X)}^2 = \left\| \sum_{j=0}^{n} f_j\xi_j \right\|_{\mathcal{M}}^2$$

for every choice of ξ and $\xi_0, \ldots, \xi_n$ in $\mathbb{C}^p$, where the f_j and $\mathcal{M}$ are as in Theorem 2.3.

The rest of the proof is divided into steps.

STEP 1. *If f_j is defined as in (2.5), then*

$$X(\alpha)f_j(\alpha)\xi = (R_0^{*j}\Lambda_0)(\alpha)\xi \ , \qquad j = 0, \ldots, n \ , \tag{12.2}$$

for every point $\alpha \in \Omega_X$ and vector $\xi \in \mathbb{C}^p$ in which R_0^* denotes the adjoint of R_0 in the Pontryagin space $\mathcal{B}(X)$.

PROOF OF STEP 1. It is readily checked that

$$\begin{aligned} X(\alpha)f_j(\alpha)\xi &= \alpha^j\Lambda_0(\alpha)\xi + \sum_{t=1}^{j} \alpha^{j-t}\frac{\Phi^{(t)}(0)}{t!}\xi \\ &= \{(R_0^j\Lambda_\alpha)(0)\}^*\xi \end{aligned}$$

for $\alpha \in \Omega_X$, $j = 0, \ldots, n$ and $\xi \in \mathbb{C}^p$. Next let R_0^* denote the adjoint of R_0 in $\mathcal{B}(X)$. Then $R_0^{*j}\Lambda_0\xi$ belongs to $\mathcal{B}(X)$ and for any choice of $\eta \in \mathbb{C}^p$ and $\alpha \in \Omega_X$,

$$\begin{aligned} \eta^*(R_0^{*j}\Lambda_0)(\alpha)\xi &= < R_0^{*j}\Lambda_0\xi, \Lambda_\alpha\eta >_{\mathcal{B}(X)} \\ &= < \Lambda_0\xi, R_0^j\Lambda_\alpha\eta >_{\mathcal{B}(X)} \\ &= \{\xi^* R_0^j\Lambda_\alpha(0)\eta\}^* \\ &= \eta^*\{R_0^j\Lambda_\alpha(0)\}^*\xi \ . \end{aligned}$$

Formula (12.2) drops out, upon combining the two exhibited formulas.

STEP 2.

$$< Xf_j\xi, Xf_i\eta >_{\mathcal{B}(X)} = < J_1 f_j\xi, f_i\eta >$$

for every choice of $i, j = 0, \ldots, n$ *and* ξ, η *in* $\mathbb{C}^p$.

PROOF OF STEP 2. By Step 1,

$$\begin{aligned} < Xf_j\xi, Xf_i\eta >_{\mathcal{B}(X)} &= < R_0^{*j}\Lambda_0\xi, R_0^{*i}\Lambda_0\eta >_{\mathcal{B}(X)} \\ &= < R_0^i R_0^{*j}\Lambda_0\xi, \Lambda_0\eta >_{\mathcal{B}(X)} \\ &= \eta^*(R_0^i R_0^{*j}\Lambda_0\xi)(0) \\ &= \eta^*(R_0^i Xf_j\xi)(0) \\ &= \eta^* \frac{(Xf_j)^{(i)}(0)}{i!}\xi \\ &= \eta^* 2h_{i-j}\xi \ , \qquad i, j = 0, \ldots, n \ . \end{aligned}$$

But that is the same as the right hand side. ∎

The main point of the preceding theorem in the present setting is that, with the help of Theorem 6.13 of [AD], it leads readily to the conclusion that if $\Phi = \tau_n[G]$ with $G \in \mathcal{L}_1$, then the kernel

$$\frac{\{\gamma_{nn}^{(n)}\}^{-1} - G(\lambda)\{\gamma_{00}^{(n)}\}^{-1}G(\omega)^*}{\rho_\omega(\lambda)}$$

has $\nu - \mu_-(H_n)$ *negative squares.* This follows from the isometry exhibited in Theorem 12.1, another application of Theorem 6.13 of [AD] and the fact that the identity

$$X(\lambda)\frac{J_1 - U(\lambda)J_1u(\omega)^*}{\rho_\omega(\lambda)}X(\omega)^* = X(\lambda)\frac{J_1 - \Theta_n(\lambda)(2\Delta_n)^{-1}\Theta_n(\omega)^*}{\rho_\omega(\lambda)}X(\omega)^*$$

can, with the help of (8.4), be reexpressed in the form

$$F(\lambda)\frac{\{\gamma_{nn}^{(n)}\}^{-1} - G(\lambda)\{\gamma_{00}^{(n)}\}^{-1}G(\omega)^*}{\rho_\omega(\lambda)}F(\omega)^* \ .$$

13. THE LEVINSON RECURSIONS

In this section we shall present a matrix version of the Levinson algorithm and of a number of related recursions which emerge from it. We shall assume throughout that all of the block Toeplitz Hermitian matrices $H_0, \ldots, H_n$ are invertible. The special case in which H_n is positive definite is sketched at the end.

The Levinson algorithm expresses the first and last block columns of Γ_k in terms of the first and last block columns of Γ_{k-1}. More precisely, upon setting

$$X_k = \begin{bmatrix} \gamma_{00}^{(k)} \\ \vdots \\ \gamma_{k0}^{(k)} \end{bmatrix} \quad \text{and} \quad Y_k = \begin{bmatrix} \gamma_{0k}^{(k)} \\ \vdots \\ \gamma_{kk}^{(k)} \end{bmatrix},$$

it turns out that if all the block Toeplitz Hermitian matrices $H_0, \ldots, H_n$ are invertible, then

$$X_k = \begin{bmatrix} X_{k-1} \\ 0 \end{bmatrix} \mu_k + \begin{bmatrix} 0 \\ Y_{k-1} \end{bmatrix} \nu_k \tag{13.1}$$

and

$$Y_k = \begin{bmatrix} X_{k-1} \\ 0 \end{bmatrix} \varepsilon_k + \begin{bmatrix} 0 \\ Y_{k-1} \end{bmatrix} \delta_k \tag{13.2}$$

for $k = 1, \ldots, n$, where

$$\mu_k = \{\gamma_{00}^{(k)} - \gamma_{0k}^{(k)}\{\gamma_{kk}^{(k)}\}^{-1}\gamma_{k0}^{(k)}\}^{-1}\gamma_{00}^{(k)} = \{\gamma_{00}^{(k-1)}\}^{-1}\gamma_{00}^{(k)} \tag{13.3}$$

$$\nu_k = \{\gamma_{kk}^{(k)}\}^{-1}\gamma_{k0}^{(k)}\mu_k = \{\gamma_{k-1,k-1}^{(k-1)}\}^{-1}\gamma_{k0}^{(k)} \tag{13.4}$$

$$\varepsilon_k = \{\gamma_{00}^{(k)}\}^{-1}\gamma_{0k}^{(k)}\delta_k = \{\gamma_{00}^{(k-1)}\}^{-1}\gamma_{0k}^{(k)} \tag{13.5}$$

$$\delta_k = \{\gamma_{kk}^{(k)} - \gamma_{k0}^{(k)}\{\gamma_{00}^{(k)}\}^{-1}\gamma_{0k}^{(k)}\}^{-1}\gamma_{kk}^{(k)} = \{\gamma_{k-1,k-1}^{(k-1)}\}^{-1}\gamma_{kk}^{(k)}\,. \tag{13.6}$$

THEOREM 13.1. *If $H_0, \ldots, H_n$ are invertible, then (13.1) and (13.2) hold and*

$$A_k(\lambda) = \{A_{k-1}(\lambda) + \lambda C_{k-1}(\lambda)\{\gamma_{kk}^{(k)}\}^{-1}\gamma_{k0}^{(k)}\}\{\gamma_{00}^{(k-1)}\}^{-1}\gamma_{00}^{(k)} \tag{13.7}$$

and

$$C_k(\lambda) = \{A_{k-1}(\lambda)\{\gamma_{00}^{(k)}\}^{-1}\gamma_{0k}^{(k)} + \lambda C_{k-1}(\lambda)\}\{\gamma_{k-1,k-1}^{(k-1)}\}^{-1}\gamma_{kk}^{(k)}, \tag{13.8}$$

for $k = 1, \ldots, n$.

PROOF. The proof is divided into steps.

STEP 1. *If H_k and H_{k-1} are invertible, then*

$$\sum_{i=0}^{k-1} h_{k-i}\gamma_{ij}^{(k-1)} = -\{\gamma_{kk}^{(k)}\}^{-1}\gamma_{kj}^{(k)} \tag{13.9}$$

for $j = 0, \ldots, k-1$ and

$$\sum_{i=1}^{k} h_{-i}\gamma_{i-1,j-1}^{(k-1)} = -\{\gamma_{00}^{(k)}\}^{-1}\gamma_{0j}^{(k)} \tag{13.10}$$

for $j = 1, \ldots, k$.

PROOF OF STEP 1. Formula (13.10) is obtained by first inverting the formula for H_n which appears in the middle of the proof of Lemma 4.2 with $n = k$. This leads to the representation

$$\Gamma_k = \begin{bmatrix} \psi & -\psi\beta\Gamma_{k-1} \\ -\Gamma_{k-1}\beta^*\psi & \Gamma_{k-1}\beta^*\psi\beta\Gamma_{k-1} + \Gamma_{k-1} \end{bmatrix},$$

with $\beta = [h_{-1} \ldots h_{-k}]$ and

$$\psi = \gamma_{00}^{(k)} = (h_0 - \beta\Gamma_{k-1}\beta^*)^{-1} = \{h_0 - \sum_{i,j=0}^{k-1} h_{-i-1}\gamma_{ij}^{(k-1)}h_{j+1}\}^{-1}.$$

Formula (13.10) drops out by matching the remaining entries in the top block row of Γ_k.

Similarly, to deduce (13.9), we first invert the other representation which is alluded to in Lemma 4.2:

$$H_k = \begin{bmatrix} I_p & 0 \\ \gamma\Gamma_{k-1} & I_{kp} \end{bmatrix} \begin{bmatrix} H_{k-1} & 0 \\ 0 & \{\gamma_{kk}^{(k)}\}^{-1} \end{bmatrix} \begin{bmatrix} I_p & \Gamma_{k-1}\gamma^* \\ 0 & I_{kp} \end{bmatrix}$$

with $\gamma = [h_k \ldots h_1]$ to obtain

$$\Gamma_k = \begin{bmatrix} \Gamma_{k-1} + \Gamma_{k-1}\gamma^*\varepsilon\gamma\Gamma_{k-1} & -\Gamma_{k-1}\gamma^*\varepsilon \\ -\varepsilon\gamma\Gamma_{k-1} & \varepsilon \end{bmatrix},$$

where

$$\varepsilon = \{h_0 - \sum_{i,j=0}^{k-1} h_{k-i}\gamma_{ij}^{(k-1)}h_{j-k}\}^{-1} = \gamma_{kk}^{(k)}.$$

Formula (13.9) now drops out by identifying the remaining blocks in the last block row of Γ_k.

STEP 2 *is to verify (13.1) and (13.2).*

PROOF OF STEP 2. In order to explain how formulas (13.3)-(13.6) are obtained, let us suppose first that there exist a choice of μ_k and ν_k, such that (13.1) holds. Then, upon multiplying through on the left by H_k and taking advantage of its block Toeplitz structure, it is readily seen that

$$I_p = \mu_k + \left\{\sum_{i=1}^{k} h_{-i}\gamma_{i-1,k-1}^{(k-1)}\right\}\nu_k$$

and

$$0 = \left\{\sum_{i=0}^{(k-1)} h_{k-i}\gamma_{i0}^{(k-1)}\right\}\mu_k + \nu_k.$$

But now, in view of formulas (13.9) and (13.10), this is clearly equivalent to the block system

$$\mu_k - \{\gamma_{00}^{(k)}\}^{-1}\gamma_{0k}^{(k)}\nu_k = I_p$$
$$-\{\gamma_{kk}^{(k)}\}^{-1}\gamma_{k0}^{(k)}\mu_k + \nu_k = 0$$

which in turn implies that if (13.1) holds, then μ_k and ν_k must be given by the first formulas listed in (13.3) and (13.4), respectively. It is now easy to check that conversely, if μ_k and ν_k are so given, then (13.1) holds.

A similar argument serves to explain the first entries in (13.5) and (13.6) and to verify (13.2). Finally, the second entries in (13.3)-(13.6) drop out easily by identifying the top and bottom blocks of X_k and Y_k in the now established formulas (13.1) and (13.2).

STEP 3 *is to complete the proof.*

PROOF OF STEP 3. To begin with, it follows easily from (13.1), (13.2) and the first sets of formulas in (13.3)-(13.6) that

$$A_k(\lambda) = \{A_{k-1}(\lambda) + \lambda\, C_{k-1}(\lambda)\{\gamma_{kk}^{(k)}\}^{-1}\gamma_{k0}^{(k)}\}\mu_k$$

and

$$C_k(\lambda) = \{A_{k-1}(\lambda)\{\gamma_{00}^{(k)}\}^{-1}\gamma_{0k}^{(k)} + \lambda C_{k-1}(\lambda)\}\delta_k\ .$$

The final result drops out from the second set of stated formulas in (13.3) and (13.4). ∎

LEMMA 13.1. *If $H_0, \ldots, H_n$ are all invertible, then the normalized polynomials*

$$P_k(\lambda) = A_k(\lambda)\{\gamma_{00}^{(k)}\}^{-1} \quad and \quad Q_k(\lambda) = C_k(\lambda)\{\gamma_{kk}^{(k)}\}^{-1}\ , \tag{13.11}$$

satisfy the recursions

$$P_k(\lambda) = P_{k-1}(\lambda) + \lambda Q_{k-1}(\lambda)\alpha_k \tag{13.12}$$

$$Q_k(\lambda) = P_{k-1}(\lambda)\beta_k + \lambda Q_{k-1}(\lambda) \tag{13.13}$$

for $k = 1, \ldots, n$, where

$$\alpha_k = \gamma_{k0}^{(k)}\{\gamma_{00}^{(k)}\}^{-1} \tag{13.14}$$

and

$$\beta_k = \gamma_{0k}^{(k)}\{\gamma_{kk}^{(k)}\}^{-1}\ . \tag{13.15}$$

PROOF. It follows readily from Theorem 13.1 that formulas (13.12) and (13.13) hold, but with

$$\alpha_k = \gamma_{k-1,k-1}^{(k-1)}\{\gamma_{kk}^{(k)}\}^{-1}\gamma_{k0}^{((k))}\{\gamma_{00}^{(k-1)}\}^{-1} \tag{13.16}$$

$$\beta_k = \gamma_{00}^{(k-1)}\{\gamma_{00}^{(k)}\}^{-1}\gamma_{0k}^{(k)}\{\gamma_{k-1,k-1}^{(k-1)}\}^{-1}\ . \tag{13.17}$$

But now, upon invoking the formulas for $\gamma_{00}^{(k-1)}$ and $\gamma_{k-1,k-1}^{(k-1)}$ which are embedded in (13.3) and (13.6), a fairly routine calculation shows that the right hand side of (13.16) [resp. (13.17)] is the same as the right hand side of (13.14) [resp. (13.15)]. ∎

The preceding recursions can be expressed conveniently in terms of the matrix

$$G_k = \begin{bmatrix} I_p & \alpha_k \\ \beta_k & I_p \end{bmatrix}, \qquad k = 1, \ldots, n , \tag{13.18}$$

which itself is easily seen to satisfy

$$G_k \Delta_k G_k^* = \Delta_{k-1} , \tag{13.19}$$

thanks to (13.3) and (13.6). Indeed, it is readily checked that

$$[\lambda C_k \quad A_k]\Delta_k^{-1} J_0 = [\lambda C_{k-1} \quad A_{k-1}]\Delta_{k-1}^{-1} J_0 \, G_k \begin{bmatrix} \lambda & 0 \\ 0 & I_p \end{bmatrix},$$

for $k = 1, \ldots, n$. This is a recursion for the top block rows of the matrices Θ_k which were introduced in Section 6. We next show that the same recursion applies for the full Θ_k matrix. The recursion can also be reversed since G_k is invertible, as is plain from (13.19); and can in fact be expressed as

$$G_k^{-1} = \begin{bmatrix} I_p & -\alpha_k \\ -\beta_k & I_p \end{bmatrix} \begin{bmatrix} I_p - \alpha_k\beta_k & 0 \\ 0 & I_p - \beta_k\alpha_k \end{bmatrix}^{-1}, \quad k = 1, \ldots, n , \tag{13.20}$$

when $H_0, \ldots, H_n$ are invertible.

THEOREM 13.2. *If $H_0, \ldots, H_n$ are invertible, then*

$$\Theta_k(\lambda)\Delta_k^{-1} J_0 = \Theta_{k-1}(\lambda)\Delta_{k-1}^{-1} J_0 G_k \begin{bmatrix} \lambda I_p & 0 \\ 0 & I_p \end{bmatrix}, \tag{13.21}$$

for $k = 1, \ldots, n$.

PROOF. It is convenient to let

$$P_k^o(\lambda) = A_k^o(\lambda)\{\gamma_{00}^{(k)}\}^{-1} \quad \text{and} \quad Q_k^o(\lambda) = C_k(\lambda)\{\gamma_{kk}^{(k)}\}^{-1}$$

and to recall that

$$C_k^o = \sum_{i=0}^{k} p_i(\lambda)\gamma_{ik}^{(k)} = \underline{\underline{p}}\Psi^* C_k$$

and that

$$A_k^o = 2I_p - \sum_{i=0}^{k} p_i(\lambda)\gamma_{i0}^{(k)} = 2I_p - \underline{\underline{p}}\Psi^* A_k .$$

Next, it follows from (13.12), upon premultiplying by Ψ^* and then projecting onto H_p^2, that

$$(2I_p - A_k^o)\{\gamma_{00}^{(k)}\}^{-1} = \{2I_p - A_{k-1}^o\}\{\gamma_{00}^{(k-1)}\}^{-1} + \lambda Q_{k-1}^o \alpha_k + (\Psi^* Q_{k-1})_{-1}\alpha_k ,$$

in which $(\Psi^*Q_{k-1})_{-1}$ denotes the coefficient of $e^{-i\theta}$ in the expression $\Psi(e^{i\theta})^*Q_{k-1}(e^{i\theta})$:

$$\begin{aligned}(\Psi^*Q_{k-1})_{-1} &= 2\sum_{i=1}^{k} h_{-i}\gamma_{i-1,k-1}^{(k-1)}\{\gamma_{k-1,k-1}^{(k-1)}\}^{-1} \\ &= -2\{\gamma_{00}^{(k)}\}^{-1}\gamma_{0k}^{(k)}\{\gamma_{k-1,k-1}^{(k-1)}\}^{-1} ,\end{aligned}$$

by (13.10). Therefore, since

$$\{\gamma_{00}^{(k)}\}^{-1} = \{\gamma_{00}^{(k-1)}\}^{-1} - \{\gamma_{00}^{(k)}\}^{-1}\gamma_{0k}\{\gamma_{k-1,k-1}^{(k-1)}\}^{-1}\alpha_k ,$$

by (13.3) and (13.16), it is readily seen that

$$-P_k^o = -P_{k-1}^o + \lambda Q_{k-1}^o\alpha_k . \tag{13.22}$$

Next, since

$$2\{\gamma_{00}^{(k-1)}\}^{-1}\beta_k + (\Psi^*Q_{k-1}^o)_{-1} = 0 ,$$

a similar analysis applied to (13.13) yields the supplementary recursion

$$Q_k^o = -P_{k-1}^o\beta_k + \lambda Q_{k-1}^o . \tag{13.23}$$

The needed justification of the bottom block row of (13.21) emerges easily from (13.22) and (13.23) to complete the proof. ∎

COROLLARY. *If $H_0, \ldots, H_n$ are invertible, then*

$$\Theta_k(\lambda) = \Theta_{k-1}(\lambda)(J_0G_k^*J_0)^{-1}\begin{bmatrix} \lambda I_p & 0 \\ 0 & I_p \end{bmatrix} \tag{13.24}$$

for $k = 1, \ldots, n$.

PROOF. This is immediate from (13.21) and the observation, coming from (13.19), that

$$\Delta_{k-1}^{-1}J_0G_kJ_0\Delta_k = (J_0G_k^*J_0)^{-1} .$$ ∎

LEMMA 13.2. *If H_n is positive definite, then the matrices Δ_jJ_0 are positive definite for $j = 0, \ldots, n$ and*

$$\begin{aligned}&(\Delta_{k-1}J_0)^{-\frac{1}{2}}G_k(\Delta_kJ_0)^{\frac{1}{2}} \\ &\quad= \begin{bmatrix} I_p & \kappa_k \\ \kappa_k^* & I_p \end{bmatrix}(\Delta_{k-1}J_0)^{-\frac{1}{2}}(\Delta_kJ_0)^{\frac{1}{2}} \\ &\quad= \begin{bmatrix} I_p & \kappa_k \\ \kappa_k^* & I_p \end{bmatrix}\begin{bmatrix} (I_p - \kappa_k\kappa_k^*)^{-\frac{1}{2}} & 0 \\ 0 & (I_p - \kappa_k^*\kappa_k)^{-\frac{1}{2}} \end{bmatrix} ,\end{aligned} \tag{13.25}$$

for $k = 1, \ldots, n$, where

$$\begin{aligned}\kappa_k &= \{\gamma_{k-1,k-1}^{(k-1)}\}^{-\frac{1}{2}} \alpha_k \{\gamma_{00}^{(k-1)}\}^{\frac{1}{2}} \\ &= \{\gamma_{k-1,k-1}^{(k-1)}\}^{\frac{1}{2}} \{\gamma_{kk}^{(k)}\}^{-1} \{\gamma_{k0}^{((k))}\} \{\gamma_{00}^{(k-1)}\}^{-\frac{1}{2}} ,\end{aligned} \tag{13.26}$$

$$(I_p - \kappa_k^* \kappa_k)^{-1} = \{\gamma_{00}^{(k-1)}\}^{-\frac{1}{2}} \{\gamma_{00}^{(k)}\} \{\gamma_{00}^{(k-1)}\}^{-\frac{1}{2}} \tag{13.27}$$

and

$$(I_p - \kappa_k \kappa_k^*)^{-1} = \{\gamma_{k-1,k-1}^{(k-1)}\}^{-\frac{1}{2}} \{\gamma_{kk}^{(k)}\} \{\gamma_{k-1,k-1}^{(k-1)}\}^{-\frac{1}{2}} . \tag{13.28}$$

PROOF. The justification is fairly straightforward except that at first glance it appears that the 21 block in the first matrix on the right hand side of (13.25) is equal to

$$\{\gamma_{00}^{(k-1)}\}^{1/2} \{\gamma_{00}^{(k)}\}^{-1} \{\gamma_{0k}^{(k)}\} \{\gamma_{k-1,k-1}^{(k-1)}\}^{-1/2}$$

and so, in order to obtain a match, it remains to show that this is equal to $\boldsymbol{\kappa}_k^*$. But

$$\begin{aligned}\boldsymbol{\kappa}_k^* &= \{\gamma_{00}^{(k-1)}\}^{-1/2} \{\gamma_{0k}^{(k)}\} \{\gamma_{kk}^{(k)}\}^{-1} \{\gamma_{k-1,k-1}^{(k-1)}\}^{1/2} \\ &= \{\gamma_{00}^{(k-1)}\}^{-1/2} \{\gamma_{0k}^{(k)}\} \{\gamma_{kk}^{(k)}\}^{-1} \{\gamma_{k-1,k-1}^{(k-1)}\} \{\gamma_{k-1,k-1}^{(k-1)}\}^{-1/2} .\end{aligned}$$

Therefore, with the help of the formulas for $\gamma_{00}^{(k-1)}$ and $\gamma_{k-1,k-1}^{(k-1)}$ which are implicit in (13.3) and (13.6), respectively, it follows that

$$\boldsymbol{\kappa}_k^* = \{\gamma_{00}^{(k-1)}\}^{1/2} \{\gamma_{00}^{(k)}\}^{-1} \{\gamma_{0k}^{(k)}\} \{\gamma_{k-1,k-1}^{(k-1)}\}^{-1/2} , \tag{13.29}$$

as needed. Finally (13.27) and (13.28) drop out by straightforward computations using (13.26), (13.29) and the formulas for $\gamma_{00}^{(k-1)}$ and $\gamma_{k-1,k-1}^{(k-1)}$ which are implicit in (13.3) and (13.6), respectively. ∎

THEOREM 13.3. *If H_n is positive definite, then the normalized matrix*

$$Y_k(\lambda) = \Theta_k(\lambda)(2\Delta_k J_0)^{-\frac{1}{2}} \tag{13.30}$$

satisfies the recursion

$$Y_k(\lambda) = Y_{k-1}(\lambda) H(\kappa_k) \begin{bmatrix} \lambda I_p & 0 \\ 0 & I_p \end{bmatrix} , \tag{13.31}$$

for $k = 1, \ldots, n$, in which $H(\kappa_k)$ is the J_0 unitary constant matrix which appears on the right hand side of (13.25). It also satisfies the inequality

$$J_1 - Y_k(\omega) J_0 Y_k(\omega)^* \geq 0 \tag{13.32}$$

for every point $\omega \in \bar{\mathbb{D}}$ *with equality on the boundary.*

PROOF. The stated recursion is immediate from (13.21) and (13.25). The final equality follows readily from (2.4), (6.4) and (6.5). ∎

Finally, we remark that the recursions simplify in the scalar case because then $\gamma_{ij}^{(n)} = \gamma_{n-j,n-i}^{(n)}$.

14. FACTORIZATION, SCHUR AND CHRISTOFFEL DARBOUX

If the block Toeplitz matrices $H_0, \ldots, H_n$ are all invertible, then the matrix valued function U which is defined in (2.6) admits a factorization as a product of elementary (block) factors

$$U_i(\lambda) = I_m + (\lambda - 1)u_i(u_i^* J_1 u_i)^{-1} u_i^* J_1 \tag{14.1}$$

based on the $m \times p$ constant matrices u_i, $i = 0, \ldots, n$. Basically the Schur algorithm is a recipe for recursively generating u_j from $U_0(\lambda), \ldots, U_{j-1}(\lambda)$ and $v_0, \ldots, v_j$ starting with $u_0 = v_0$. Having obtained $U_0, \ldots, U_{j-1}$, u_j may be defined by (14.5). However, it remains to check that the constant $p \times p$ matrix $u_j^* J_1 u_j$ is invertible, so that $U_j(\lambda)$ can be defined and the algorithm can continue. The justification of this depends upon the presumed invertibility of $H_0, \ldots, H_j$. We shall first define u_j by a different recipe in Theorem 14.1 and then, in Theorem 14.2, show that this recipe is the same as (14.5).

Before proceeding to the proof it is perhaps well to motivate the indicated form of $U_i(\lambda)$ by recalling that if $n = 0$, then the space $\mathcal{M}$ considered in Theorem 2.3 consists only of the span of the column vectors of $f_0(\lambda) = v_0$. In this instance formula (2.6) with $\alpha = 1$ reduces to (14.1) with $u_i = v_0$.

THEOREM 14.1. *If the block Toeplitz matrices* $H_0, \ldots, H_n$ *are all invertible, then the space*

$$\mathcal{M}_j = \text{the span of the columns of } \{f_0, \ldots, f_j\} ,$$

$j = 0, \ldots, n$, *endowed with the* J_1 *inner product is a reproducing kernel Pontryagin space (with negative space of dimension* $\mu_-(H_j)$*) and reproducing kernel*

$$K_\omega^j(\lambda) = \frac{J_1 - W_j(\lambda) J_1 W_j(\omega)^*}{\rho_\omega(\lambda)} , \tag{14.2}$$

where

$$W_j(\lambda) = U_0(\lambda) \ldots U_j(\lambda) ,$$

and $U_i(\lambda)$ *is given by (14.1) with*

$$u_0 = v_0 ,$$

$$u_i = \{W_{i-1}(\lambda)\}^{-1} \sum_{s=0}^{i} f_s(\lambda) \gamma_{si}^{(i)} \{\gamma_{ii}^{(i)}\}^{-1} , \qquad i = 1, \ldots, n , \tag{14.3}$$

and

$$u_i^* J_1 u_i = 2\{\gamma_{ii}^{(i)}\}^{-1} , \qquad i = 0, \ldots, n . \tag{14.4}$$

PROOF. To begin with it is readily checked that U_0 is well defined, since $u_0^* J_1 u_0 = v_0^* J_1 v_0$ is invertible, and that $K_\omega^0(\lambda)$ is a reproducing kernel for $\mathcal{M}_0$. Thus u_1 can be defined via (14.3) and it is not too hard to check directly that u_1 is a constant vector which meets (14.4) and hence that $U_1(\lambda)$ is well defined and that $K_\omega^1(\lambda)$ is the reproducing kernel for $\mathcal{M}_1$.

Now let us proceed by induction, supposing that we have successfully defined $U_0, \ldots, U_{j-1}$ and identified $K_\omega^{j-1}(\lambda)$ as the reproducing kernel for $\mathcal{M}_{j-1}$. To this end, let

$$g_j(\lambda) = \sum_{s=0}^{j} f_s(\lambda)\gamma_{sj}^{(j)}\{\gamma_{jj}^{(j)}\}^{-1} .$$

Then it is readily checked that

$$< J_1 f_i, g_j > = \begin{cases} 0 & \text{if } i = 0, \ldots, j-1 \\ 2\{\gamma_{jj}^{(j)}\}^{-1} & \text{if } i = j , \end{cases}$$

and hence that the columns of g_j are orthogonal to $\mathcal{M}_{j-1}$ and yet span out the J_1 orthogonal complement of $\mathcal{M}_{j-1}$ in $\mathcal{M}_j$. Thus the reproducing kernel for $\mathcal{M}_j$,

$$\begin{aligned} K_\omega^j(\lambda) &= K_\omega^{j-1}(\lambda) + g_j(\lambda)\{< J_1 g_j, g_j >\}^{-1} g_j(\omega)^* \\ &= K_\omega^{j-1}(\lambda) + \frac{g_j(\lambda)\gamma_{jj}^{(j)} g_j(\omega)^*}{2} . \end{aligned}$$

The next step is to verify that $\{W_{j-1}(\lambda)\}^{-1} g_j(\lambda)$ is a constant matrix. In view of the calculations already carried out in Section 13 this is accomplished most easily by recognizing that

$$W_{j-1}(\lambda) = \Theta_{j-1}(\lambda) N_{j-1}$$

for some suitably chosen constant matrix N_{j-1}, as follows from the fact that the reproducing kernel for $\mathcal{M}_{j-1}$ can also be written as

$$K_\omega^{j-1}(\lambda) = \frac{J_1 - \Theta_{j-1}(\lambda)(2\Delta_{j-1})^{-1}\Theta_{j-1}(\omega)^*}{\rho_\omega(\lambda)}$$

and there is only such. At the same time since

$$g_j(\lambda) = \begin{bmatrix} C_j(\lambda) \\ C_j^o(\lambda) \end{bmatrix} \{\gamma_{jj}^{(j)}\}^{-1} ,$$

it now follows easily from (13.21) that u_j is indeed a constant vector. The evaluation (14.4) then drops out easily by using (14.3) to compute

$$u_j^* J_1 u_j = < J_1 W_{j-1} u_j, W_{j-1} u_j >$$

another way. Thus, putting it all together, we obtain

$$\begin{aligned} K_\omega^j(\lambda) &= K_\omega^{j-1}(\lambda) + W_{j-1}(\lambda)u_j(u_j^* J_1 u_j)^{-1}u_j^* W_{j-1}(\omega)^* \\ &= K_\omega^{j-1}(\lambda) + W_{j-1}(\lambda)\left\{\frac{J_1 - U_j(\lambda)J_1U_j(\omega)^*}{\rho_\omega(\lambda)}\right\}W_{j-1}(\omega)^* \\ &= \frac{J_1 - W_j(\lambda)J_1W_j(\omega)^*}{\rho_\omega(\lambda)} , \end{aligned}$$

with $W_j = W_{j-1}U_j$, as claimed. ∎

THEOREM 14.2. *If $H_0, \ldots, H_n$ are invertible, then the $m \times p$ constant matrices u_j given by (14.3), for $j = 1, \ldots, n$, may also be obtained via the recipe*

$$u_j^* J_1 = \lim_{\lambda \to 0} \frac{v_0^* + \lambda v_1^* + \ldots + \lambda^j v_j^*}{\lambda^j} J_1 U_0(\lambda) \ldots U_{j-1}(\lambda) . \tag{14.5}$$

PROOF. The formula is readily established for $j = 1$. Suppose further that it is known to be valid for $j = 1, \ldots, k-1$. Then

$$\begin{aligned} &\lim_{\lambda \to 0} \frac{v_0^* + \lambda v_1^* + \ldots + \lambda^t v_t^*}{\lambda^t} J_1 U_0(\lambda) \ldots U_{k-1}(\lambda) \\ &\quad = u_t^* J_1 U_t(0) \ldots U_{k-1}(0) \\ &\quad = 0 \end{aligned}$$

for $t = 0, \ldots, k-1$. But this is the same as to say that

$$\lim_{\lambda \to 0} f_t^{\#}(\lambda) J_1 W_{k-1}(\lambda) = 0$$

for $t = 0, \ldots, k-1$, or equivalently, since

$$\{W_{k-1}(\lambda)\}^{-1} = J_1 W_{k-1}^{\#}(\lambda) J_1 , \qquad \lambda \neq 0 ,$$

that

$$\lim_{\lambda \to \infty} \{W_{k-1}(\lambda)\}^{-1} f_t(\lambda) = 0$$

for $t = 0, \ldots, k-1$. Thus, by (14.3),

$$\begin{aligned} u_k &= \{W_{k-1}(\lambda)\}^{-1} \sum_{t=0}^{k} f_t(\lambda) \gamma_{tk}^{(k)} \{\gamma_{kk}^{(k)}\}^{-1} \\ &= \lim_{\lambda \to \infty} \{W_{k-1}(\lambda)\}^{-1} \sum_{t=0}^{k} f_t(\lambda) \gamma_{tk}^{(k)} \{\gamma_{kk}^{(k)}\}^{-1} \\ &= \lim_{\lambda \to \infty} \{W_{k-1}(\lambda)\}^{-1} f_k(\lambda) , \end{aligned}$$

which is equivalent to (14.5) for $j = k$. ∎

We remark that the fact that

$$< J_1 g_i, g_j > = \begin{cases} 0 & i \neq j \\ 2\{\gamma_{jj}^{(j)}\}^{-1} & i = j \end{cases}$$

when $H_0, \ldots, H_n$ are invertible and

$$g_j = \sum_{t=0}^{j} f_t(\lambda)\gamma_{tj}^{(j)}\{\gamma_{jj}^{(j)}\}^{-1} , \qquad j = 0, \ldots, n ,$$

is equivalent to the triangular factorization formula

$$V_n^* H_n V_n = \operatorname{diag}(\{\gamma_{00}^{(0)}\}^{-1}, \{\gamma_{11}^{(1)}\}^{-1}, \ldots, \{\gamma_{nn}^{(n)}\}^{-1}) , \tag{14.6}$$

where V_n is upper block triangular matrix with

$$(V_n)_{ij} = \gamma_{ij}^{(j)}\{\gamma_{jj}^{(j)}\}^{-1} , \qquad 0 \leq i \leq j \leq n .$$

The analysis furnished in the proof of Theorem 14.1 also serves to exhibit

$$\mathcal{M}_j = \mathcal{N}_0 \boxplus \mathcal{N}_1 \boxplus \ldots \boxplus \mathcal{N}_j ; \qquad j = 0, \ldots, n ,$$

as an orthogonal direct sum decomposition of the reproducing kernel Pontryagin spaces $\mathcal{N}_j$ which are spanned by the columns of

$$g_j(\lambda) = \begin{bmatrix} C_j(\lambda) \\ C_j^o(\lambda) \end{bmatrix} \{\gamma_{jj}^{(j)}\}^{-1}$$

with reproducing kernel

$$g_j(\lambda)\{< J_1 g_j, g_j >\}^{-1} g_j(\omega)^* = g_j(\lambda)\gamma_{jj}^{(j)} g_j(\omega)^*/2 .$$

THEOREM 14.3. *If $H_0, \ldots, H_n$ are all invertible, then*

$$\frac{J_1 - \Theta_n(\lambda)(2\Delta_n)^{-1}\Theta_n(\omega)^*}{\rho_\omega(\lambda)} = \sum_{j=0}^{n} \begin{bmatrix} C_j(\lambda) \\ C_j^o(\lambda) \end{bmatrix} \{2\gamma_{jj}^{(j)}\}^{-1} \begin{bmatrix} C_j(\omega) \\ C_j^o(\omega) \end{bmatrix}^* . \tag{14.7}$$

PROOF. You have only to match reproducing kernels of $\mathcal{M}_n$. ∎

The familiar Christoffel Darboux formula

$$\frac{A_n(\lambda)\{\gamma_{00}^{(n)}\}^{-1}A_n(\omega)^* - \lambda C_n(\lambda)\{\gamma_{nn}^{(n)}\}^{-1}C_n(\omega)^*\omega^*}{\rho_\omega(\lambda)} = \sum_{j=0}^{n} C_j(\lambda)\{\gamma_{jj}^{(j)}\}^{-1}C_j(\omega)^* \tag{14.8}$$

is just the 11 block of (14.7). There is of course an analogous formula for polynomials of the second kind which is obtained from the 22 block of (14.7).

Finally we remark that, for suitably restricted interpolants Φ, the matrix valued function

$$\begin{aligned} X(\lambda) &= [\Phi(\lambda) \ \ I_p] \\ &= v_0^* + \lambda v_1^* + \ldots + \lambda^n v_n^* + O(\lambda^{n+1}) \end{aligned}$$

generates a reproducing kernel Pontryagin space $\mathcal{B}(X)$ with reproducing kernel

$$\frac{X(\lambda)J_1X(\omega)^*}{\rho_\omega(\lambda)} = \frac{\Phi(\lambda)+\Phi(\omega)^*}{\rho_\omega(\lambda)} \ .$$

The preceding analysis can be adapted to show that

$$\mathcal{B}(X) = \mathcal{B}(XU_0 \ldots U_j) \boxplus X\mathcal{M}_j \ .$$

For more information on such decompositions and the corresponding reproducing kernel space interpretation of the Schur algorithm, see [AD].

15. REFERENCES

[AD] D. Alpay and H. Dym, *On applications of reproducing kernel spaces to the Schur algorithm and rational J unitary factorization,* in: I. Schur Methods in Operator Theory and Signal Processing, (I. Gohberg, ed.), Operator Theory: Advances and Applications, **OT18**, Birkhäuser Verlag, Basel, 1986, pp. 89-159.

[AG] Alpay, D. and I. Gohberg, *On orthogonal matrix polynomials,* this issue.

[CG] Clancey, K. and I. Gohberg, *Factorization of matrix functions and singular integral operators,* Operator Theory Advances and Applications, **OT3**, Birkhäuser Verlag, Basel, 1981.

[dB] L. de Branges, *Some Hilbert spaces of analytic functions I,* Trans. Amer. Math. Soc. **106** (1963), 445–468.

[D] Dym, H., *J Contractive Matrix Functions, Reproducing Kernel Hilbert Spaces and Interpolation,* CBMS Lecture Notes, in press.

[DGK1] Delsarte, Ph., Y. Genin and Y. Kamp, *Orthogonal polynomial matrices on the unit circle,* IEEE Trans. Circuits and Systems, **25** (1978), 145-160.

[DGK2] Delsarte, Ph., Y. Genin and Y. Kamp, *Pseudo-Caratheodory functions and Hermitian Toeplitz matrices,* Philips J. Research **41** (1986), 1-54.

[F] Fuhrmann, P.A., *Orthogonal matrix polynomials and system theory,* Preprint, September 1986.

[GH] Gohberg, I.C. and G. Heinig, *Inversion of finite Toeplitz matrices with entries being elements from a noncommutative algebra,* Rev. Roumaine Math. Pures Appl. **19** (1974), 623-665.

[H1] Hirschman, I.I., Jr., *Matrix valued Toeplitz operators*, Duke Math. J., **34** (1967), 403-416.

[H2] Hirschman, I.I., Jr., *Recent developments in the theory of finite Toeplitz operators*, Advances in Probability - I, (P. Ney, ed.), Marcel Dekker, New York, 1972, pp. 103-167.

[K1] Krein, M.G. *Continuous analogs of theorems on polynomials orthogonal on the unit circle*, Dokl. Akad. Nauk SSSR **105** (1955), 637-640.

[K1] Krein, M.G., *Distribution of roots of polynomials orthogonal on the unit circle with respect to a sign alternating weight*, Teor. Funkcii Funkcional Anal. i Prilozen **2** (1966), 131-137 (Russian).

[Ka] Kailath, T., *A view of three decades of linear filtering theory*, IEEE Trans. Information Theory **20** (1974), 145-181.

[KL] Krein, M.G. and H. Langer, *On some continuation problems which are closely related to the theory of operators in spaces* Π_x*. IV: Continuous analogues of orthogonal polynomials on the unit circle with respect to an indefinite weight and related continuation problems for some classes of functions*, J. Oper. Theory **13** (1985), 299-417.

[KVM] Kailath, T., A. Vieira and M. Morf, *Inverses of Toeplitz operators, innovations, and orthogonal polynomials*, SIAM Review **20** (1978), 106-119.

[LT] Lancaster, P. and M. Tismenetsky, *The Theory of Matrices*, Second Edition, Academic Press, Orlando, 1985.

[MVK] Morf, M., A. Vieira and T. Kailath, *Covariance characterization by partial autocorrelation matrices*, Annals of Statistics **3** (1978), 643-648.

[Y] Youla, D.C., *Interpolatory multichannel spectral estimation*, I. *General theory and the FEE*, Preprint, July 1979.

[YK] Youla, D.C. and N.N. Kazanjian, *Bauer type factorization of positive matrices and the theory of matrix polynomials orthogonal on the unit circle*, IEEE Trans. Circuits Syst. **25** (1978), 57-69.

Department of Theoretical Mathematics
The Weizmann Institute of Science
Rehovot 76100, Israel

Operator Theory:
Advances and Applications, Vol. 34

MATRIX GENERALIZATIONS OF M.G. KREIN THEOREMS ON ORTHOGONAL POLYNOMIALS

I. Gohberg and L. Lerer

The results of M.G. Krein regarding polynomials that are orthogonal on the unit circle with respect to a sign alternating weight function, are generalized to the case of matrix polynomials. These results are concerned with the distribution of the zeroes of the orthogonal polynomials and with the inverse problem of reconstructing the weight function from a given polynomial.

TABLE OF CONTENTS

0. INTRODUCTION

In this paper we generalize the following two theorems of M.G. Krein [20].

THEOREM 0.1. *Let $t_{-n},\dots,t_{-1},t_0,t_1,\dots,t_n$ be complex numbers such that the matrices $T_k = [t_{p-q}]_{p,q=0}^{k}$ $(k=0,1,\dots,n)$ are hermitian and non-singular, and let β (respectively, γ) stand for the number of constancies (respectively, alterations) of sign in the sequence*

$$1,\ D_0,\ D_1,\ \dots,\ D_{n-1} \qquad \left(D_k := \det T_k\right).$$

Let $f_n(z) = x_0 z^n + x_1 z^{n-1} + \dots + x_n$ *be a polynomial whose coefficients are found from the equation*

$$T_n \begin{bmatrix} x_0 \\ x_1 \\ \vdots \\ x_n \end{bmatrix} = \begin{bmatrix} 1 \\ 0 \\ \vdots \\ 0 \end{bmatrix} . \tag{0.1}$$

If $D_n D_{n-1} > 0$ *(respectively,* $D_n D_{n-1} < 0$*), then the polynomial* f_n *has* β *(resp.,* γ*) zeroes inside the unit circle and* γ *(resp.,* β*) zeroes outside the unit circle (the zeroes are counted with multiplicities).*

THEOREM 0.2. *For a polynomial* $f_n(z) = x_0 z^n + x_1 z^{n-1} + \dots + x_n$, *with* $x_0 \neq 0$, *there exists a hermitian invertible Toeplitz matrix* $T_n = [t_{j-k}]_{j,k=0}^{n}$ *such that (0.1) holds true if and only if* x_0 *is real and* f_n *has no zeroes on the unit circle and no pair of zeroes that are symmetric with respect to the unit circle.*

Note that the polynomial $f_n(z)$ in the above theorems is the nth orthogonal polynomial in the indefinite inner product space of complex polynomials with inner product

$$\langle f,g \rangle := \int_{-\pi}^{\pi} f(e^{i\theta})\overline{g(e^{i\theta})}h(e^{i\theta})d\theta , \tag{0.2}$$

where the weight $h(z)$ is a real-valued Lebesgue-integrable function on the unit circle, whose Fourier coefficients

$$h_k := \frac{1}{2\pi} \int_{-\pi}^{\pi} h(e^{i\theta})e^{-ik\theta}d\theta$$

coincide with t_k for $k = 0,\pm1,\dots,\pm n$. For full proofs of the above theorems, as well as further refinements and references, see the paper [8]. We note only that the invertibility of the matrices T_k $(k=0,1,\dots,n)$ ensures that the orthogonalization process can indeed be applied to the sequence $1, z, \ldots, z^n$ (see e.g., [14], Section 1.2), and the polynomial $f_n(z) = \sum_{j=0}^{n} x_{n-j} z^j$, determined by the equation (0.1), is indeed the (n+1) element in the orthogonalized sequence of polynomials.

Our generalizations are concerned with the case when t_j $(j=0, \pm 1, \ldots \pm n)$ are rxr complex matrices and when $x_0, x_1, \ldots, x_n$ are also rxr complex matrices, which are solutions of the equation(0.1), where $T_n = [t_{j-k}]_{j,k=0}^{n}$ is a block Toeplitz matrix, and the number 1 in the right hand side is replaced by the rxr identity matrix I. For this matrix case, Theorem 0.1 admits a natural generalization. Namely, we show that if the matrix x_0 is positive definite, then the number of zeroes of the polynomial $\det(z^n x_0 + z^{n-1} x_1 + \ldots + x_n)$ inside the unit circle (counting multiplicities) is equal to the number of positive eigenvalues of the nrxnr block Toeplitz matrix $T_{n-1} = [t_{j-k}]_{j,k=0}^{n-1}$ and there are no zeroes of the above polynomial on the unit circle. The condition $x_0 > 0$ is essential here, and simple examples show that if x_0 is not definite, then $\det(z^n x_0 + z^{n-1} x_1 + \ldots + x_n)$ may have zeroes even on the unit circle.

The second Krein theorem does not admit a generalization which can be stated in terms of the zeroes of the polynomial $\det(z^n x_0 + z^{n-1} x_1 + \ldots + x_n)$ only. It turns out that the equation

$$\det\left(\sum_{j=0}^{n} z^j x_{n-j} \right) = 0$$

may have pairs of solutions that are symmetric with respect to the unit circle, but in such case the corresponding chains of generalized eigenvectors of the matrix polynomial $\sum_{j=0}^{n} z^j x_{n-j}$ are subject to some additional geometric conditions of orthogonality nature (see Theorems 6.2, 8.1).

Note that Theorem 0.2 and its matrix generalization deal with a specific inverse problem for (block) Toeplitz matrices, which is to determine a (block) Toeplitz invertible hermitian matrix T_n given the first (block) column $x = col(x_j)_{j=0}^n$ of its inverse T_n^{-1}. In this paper we present two different approaches for solving this problem.

The starting point for the first approach is the paper [10] by I. Gohberg and G. Heinig, where they solve the basic inverse problem of determining an invertible block Toeplitz matrix T via the first and the last block columns and block rows of the inverse T^{-1}. Based on some recent developments in the spectral theory of matrix polynomials (see [11],[12], [13]), we modify the results of [10] to a form which allows reducing the above mentioned specific inverse problem to a certain factorization problem. In the case $x_0 > 0$ this factorization problem reads as follows: For the rxr matrix function

$$F(\lambda) := [A(\lambda)]^* A(\lambda), \quad (|\lambda| = 1), \tag{0.3}$$

where $A(\lambda) = x_0^{-\frac{1}{2}} \sum_{j=0}^{n} \lambda^j x_{n-j}$, find a factorization of the form

$$F(\lambda) := B(\lambda)[B(\lambda)]^*, \quad (|\lambda| = 1), \tag{0.4}$$

where $B(\lambda) = \sum_{j=0}^{n} \lambda^j b_j$ is a rxr matrix polynomial of degree n with positive definite leading coefficient y_n, such that $A(\lambda)$ and $[B(\lambda)]^*$ are right coprime, i.e.,

$$\mathrm{Ker}A(z) \cap \mathrm{Ker}[B(z)]^* = (0) \quad (z \in \mathbb{C}). \tag{0.5}$$

In the particular case when all the zeroes of detA(z) are inside the unit circle, (0.3) can be modified to a left Wiener-Hopf factorization with respect to the unit circle of $F(\lambda)$: $F(\lambda) = [F_\ell(\lambda)]^* F_\ell(\lambda)$, $(|\lambda|=1)$, where $F_\ell(\lambda) := \lambda^{-n} A(\lambda)$. Then well known results (see e.g., [5],[14]) ensure that $F(\lambda)$ admits a right Wiener-Hopf factorization $F(\lambda) = F_r(\lambda)[F_r(\lambda)]^*$ with respect to the unit circle such that $F_r(\lambda) = \lambda^{-n} B(\lambda)$, where $B(\lambda) = \sum_{j=0}^{n} \lambda^j b_j$ is a rxr matrix polynomial with a positive definite leading coefficient, and all the zeroes of $B(\lambda)$ are inside the unit circle. Clearly this polynomial $B(\lambda)$ satisfies (0.4) and (0.5).

In this paper we solve the above stated factorization problem in its general setting. An important ingredient in the analysis of this factorization problem is a criterium for existence matrix polynomial solutions $X(\lambda)$, $Y(\lambda)$ of the equation

$$M(\lambda)X(\lambda) + Y(\lambda)L(\lambda) = R , \tag{0.6}$$

where $M(\lambda)$ and $L(\lambda)$ are given $r \times r$ matrix polynomials and $R \in \mathbb{C}^{r\times r}$. Here some unpublished material of the present authors [16] (see also [15]) as well as the results of L. Lerer and M. Tismenetsky [24], [25] played an essential role.

The above approach to the matrix generalization of Theorem 0.2 gives a better understanding of its spectral nature. However, the algorithm which this approach provides for reconstructing the matrix T_n from the first block column of T_n^{-1}, is rather complicated.

The second approach presented in this paper provides a transparent construction of the matrix T_n. We start developing this approach by considering a more general problem of determining a block Toeplitz matrix T_n (which is not required to be invertible) that satisfies (0.1) and the equation

$$[v_0 v_1 \dots v_n] T_n = [I \ 0 \dots 0] . \tag{0.7}$$

It turns out that such matrix T_n exists if and only if the following matrix equation of Stein type is solvable:

$$S - \hat{K}_{\tilde{V}^0} S K_{\tilde{X}^0} = \begin{bmatrix} x_0^{-1} & 0 & \cdots & 0 \\ 0 & 0 & & \vdots \\ \vdots & \vdots & & \\ 0 & 0 & \cdots & 0 \end{bmatrix} , \tag{0.8}$$

where $\hat{K}_A$ and K_A are companion type matrices (see (2.2)) and

$$\tilde{v}^0(\lambda) := \sum_{j=0}^{n} \lambda^j v_0^{-1} v_{n-j}, \quad \tilde{x}^0(\lambda) = \sum_{j=0}^{n} \lambda^j x_{n-j} x_0^{-1}$$

(see [21] where this fact is stated without proof in a somewhat different form). Moreover, the desired matrix T_n is expressed via the solutions S of (0.8) as follows:

$$T_n = \left[\begin{array}{c|ccc} v_0^{-1} & -v_0^{-1}v_1 & \cdots & -v_0^{-1}v_n \\ \hline & I & & \\ 0 & & \ddots & \\ & & & I \end{array}\right] \left[\begin{array}{c|c} x_0 & 0 \\ \hline 0 & S \end{array}\right] \left[\begin{array}{c|ccc} x_0^{-1} & & 0 & \\ \hline -x_1 x_0^{-1} & I & & \\ \vdots & & \ddots & \\ -x_n x_0^{-1} & & & I \end{array}\right]. \tag{0.9}$$

The next step is the analysis of conditions that ensure solvability of the equation (0.8). Here again we use the criterium for solvability of equations of the type (0.6) and the results of [24]-[25] about connections between equations of type (0.6) and Lyaponov type matrix equations.

Finally, specifying the solution of the above stated general problem for the case when $x_0 > 0$ and T_n is required to be hermitian, we obtain the matrix generalization of Theorem 0.2. A formula for T_n in this case is obtained by setting $v_j = x_j^*$ in (0.8) and (0.9).

The topics discussed in this paper have direct relation to inverse spectral problems for rational matrix functions, especially for those that are unitary on the unit circle or unitary in an indefinite metric. One may find some of these connections in the paper [1]. We are planning to dedicate a separate publication to this topic, where we shall also consider the case of a non-definite coefficient x_0.

The analysis of matrix orthogonal polynomials was the main topic of the Operator Theory Seminar held at Tel-Aviv University during the second semester of the 1986-87 academic year. Parallel to the present paper, three other papers concerning this topic were written. We have in mind the paper by D. Alpay and I. Gohberg [1], which contains, in particular, the matrix generalization of Theorem 0.1; the paper of A. Atzmon [2], where an operator generalization of Theorem 0.2 is obtained in terms of solvability of an equation of type (0.6); and the paper by A. Ben-Artzi and I. Gohberg [4], which is concerned with non-stationary generalizations of Theorem 0.1 for block matrices. All these papers are based on different methods and ideas.

The rest of the paper is organized as follows. The first section is of preliminary character. The matrix generalization of Theorem 0.1 is established in Section 2. In the third section we present our modification of the I.Gohberg and G. Heinig [10] results concerning the basic inverse problem for invertible block Toeplitz matrices. Section 4 contains criteria for solvability of equations of type (0.6) and (0.8). The analysis of the factorization problem determined by (0.3)-(0.5) is given in Section 5. Section 6 contains the first proof of the matrix generalization of Theorem 0.2. In Section 7 we discuss inverse problems for general block Toeplitz matrices (that are not required to be invertible) and their connection to matrix equations of Stein type. Using the results of Section 7 we give, in Section 8, the second proof of the matrix generalization of Theorem 0.2. Note that this proof and Section 7 are independent of Sections 3, 5 and 6. Also, in Section 8, we present formulas for the factors in a right symmetric factorization (0.4) that is relatively coprime (i.e., (0.5) holds true) with a given left symmetric factorization (0.3).

1. PRELIMINARIES

Throughout this paper all matrices are assumed to have complex entries. When convenient we shall indentify an n×m matrix A with its canonical representation as a linear transformation from $\mathbb{C}^m$ into $\mathbb{C}^n$.

The standard inner product in $\mathbb{C}^n$ is denoted by < , >. The superscript "*" (as in A^*) stands for the adjoint matrix or operator. We shall also use the superscript "T" (as in A^T) to denote the transposed matrix. Direct sum of two subspaces M and N in $\mathbb{C}^n$ is denoted by $M \dot{+} N$.

We shall use the following notations concerning block matrices. The one-column block matrix whose i^{th} entry is equal to the matrix A_i (i=1,...,k), will be denoted by $\operatorname{col}(A_i)_{i=1}^k$. Similarly, row $(A_i)_{i=1}^k$ denotes the one-row block matrix $[A_1 A_2 \dots A_k]$. The symbol $A_1 \dot{+} A_2 \dot{+} \dots \dot{+} A_k$ as well as $\operatorname{diag}(A_1,\dots,A_k)$ will be used to denote the diagonal block matrix $[\delta_{ij}A_j]_{i,j=1}^k$, where δ_{ij} is the Kronecker delta. If $D = \operatorname{diag}(d,d,\dots,d)$ and $B = [b_{ij}]_{i,j=1}^k$ $(d, b_{ij} \in \mathbb{C}^{r\times r})$, we shall use the notations

$$dB := DB , \qquad Bd := BD.$$

A pair of complex matrices (Φ,Q) is referred to as a *right admissible pair* of order p if Φ is of size r×p and Q is of size p×p (so one can form matrices of type ΦQ^k). The number r will be fixed throughout the paper, while p may depend on the admissible pair. A pair (R,Ψ) with $R \in \mathbb{C}^{p\times p}$, $\Psi \in \mathbb{C}^{p\times r}$ is called a *left admissible pair* of order p. Note that the pairs we consider are assumed to be right admissible, if not specified otherwise. The notions below can be reformulated for left admissible pairs in an obvious way.

For a pair (Φ,Q) we define the number

$$\operatorname{ind}(\Phi,Q) := \min\{m \mid \operatorname{Ker} \operatorname{col}(\Phi Q^{i-1})_{i=1}^m = \operatorname{Ker} \operatorname{col}(\Phi Q^{i-1})_{i=1}^{m+1}\},$$

which is called the *index of stabilization* of the pair (Φ,Q). Since the subspaces $\operatorname{Ker} \operatorname{col}(\Phi Q^{i-1})_{i=1}^m$ (m=1,2,...) form a descending chain (by inclusion), the index of stabilization $\operatorname{ind}(\Phi,Q)$ always exists because of the finite dimensionality of $\mathbb{C}^p$. We shall denote

$$\mathrm{Ker}(Q,U) := \bigcap_{j=1}^{\infty} \ker \Phi Q^{j-1} = \mathrm{Ker}\ \mathrm{col}(\Phi Q^{j-1})_{j=1}^{s},$$

where $s = \mathrm{ind}(\Phi,Q)$.

Two pairs (Φ_1,Q_1) and (Φ_2,Q_2) of order p are called *similar* if there is a pxp invertible matrix S such that $\Phi_1 = \Phi_1 S$ and $Q_1 = S^{-1}Q_2 S$. Clearly, similar pairs have the same index of stabilization.

Passing to matrix polynomials we need the following. Given a sequence $\ell_0, \ell_1, \ldots, \ell_n$ of rxr complex matrices, we shall always use the corresponding capital letter L for the expression

$$L(\lambda) = \sum_{j=0}^{n} \lambda^j \ell_j , \tag{1.1}$$

where λ is a complex variable. The expression (1.1) is called a *matrix polynomial of degree* ν, and in this case the matrix ℓ_n is referred to as the *leading coefficient* of $L(\lambda)$ (we do not exclude the case $\ell_n = 0$). Of course, for any integer k, k>n, the expression (1.1) can also be viewed as a matrix polynomial of degree k (with zero leading coefficient). If the matrix ℓ_n in (1.1) is non-singular, we say that $L(\lambda)$ is a *matrix polynomial of degree* n *with invertible leading coefficient*. A polynomial of this kind with $\ell_n = I$, the rxr identity matrix, will be referred to as a *monic matrix polynomial of degree* n. If the matrix ℓ_0 is non-singular, $L(\lambda)$ is called a *comonic matrix polynomial*. If $L(\lambda)$ is a matrix polynomial of degree n defined by (1.1), we shall always denote

$$L^{O}(\lambda) := \lambda^n L(\lambda^{-1}) = \sum_{j=0}^{n} \lambda^j \ell_{n-j}. \tag{1.2}$$

Clearly, if $L(\lambda)$ is a comonic polynomial, then $L^{O}(\lambda)$ is monic.

We now recall some basic facts from the spectral theory of matrix polynomials (see the monograph [13] for a detailed exposition). The point $\lambda_0 \in \mathbb{C}$ is an *eigenvalue* of the matrix polynomial $L(\lambda)$ if $\det L(\lambda_0) = 0$. The set of all eigenvalues of $L(\lambda)$ is called the *spectrum* of $L(\lambda)$, and is denoted by $\sigma(L)$. The polynomial $L(\lambda)$ is said to be *regular* if $\sigma(L) \neq \mathbb{C}$. In this case the spectrum $\sigma(L)$ is either a finite set or else it is empty. Clearly, monic and comonic matrix polynomials are regular. If $\lambda_0 \in \sigma(L)$, then any non-zero (column) vector in $\operatorname{Ker}L(\lambda_0)$ is called a *right eigenvector* of $L(\lambda)$ corresponding to λ_0.

Let $L(\lambda)$ be a regular $r\times r$ matrix polynomial, $\lambda_0 \in \sigma(L)$ and $s = \dim\operatorname{Ker}L(\lambda_0)$. The *local Smith form* of L at λ_0 is defined as the representation

$$L(\lambda) = E_{\lambda_0}(\lambda)D_{\lambda_0}(\lambda)F_{\lambda_0}(\lambda) , \tag{1.3}$$

where $E_{\lambda_0}(\lambda)$, $F_{\lambda_0}(\lambda)$ are matrix polynomials with $\det E_{\lambda_0}(\lambda_0) \neq 0 \neq \det F_{\lambda_0}(\lambda_0)$,

$$D_{\lambda_0}(\lambda) = \operatorname{diag}\left((\lambda-\lambda_0)^{\nu_1},(\lambda-\lambda_0)^{\nu_2},\ldots,(\lambda-\lambda_0)^{\nu_s},1,\ldots,1\right)$$

and $\nu_1 \geqslant \nu_2 \geqslant \ldots \geqslant \nu_s \geqslant 1$ are integers called *partial multiplicities* of $L(\lambda)$ at λ_0.

A sequence of r-dimensional (column) vectors $\phi_0,\phi_1,\ldots,\phi_k$ ($\phi_0 \neq 0$) is called a *right Jordan chain of length* $k+1$ of $L(\lambda)$ corresponding to $\lambda_0 \in \sigma(L)$ if the equalities

$$\sum_{p=0}^{i} \frac{1}{p!} L^{(p)}(\lambda_0)\phi_{i-p} = 0 \qquad (i = 0,1,\ldots,k)$$

hold true. Here $L^{(p)}(\lambda)$ denotes the pth derivative of $L(\lambda)$ with respect to λ. Note that the leading vector ϕ_0 of a Jordan chain is an eigenvector of L corresponding to λ_0.

It turns out that one can find a basis $\phi_{10},\phi_{20},\ldots,\phi_{s0}$ of the subspace $\operatorname{Ker}L(\lambda_0)$ such that each vector ϕ_{i0} is a leading vector of a right Jordan chain of length ν_i

$$\phi_{i0}, \phi_{i1}, \ldots, \phi_{i,\nu_i-1} \quad (i = 1, \ldots, s) , \tag{1.4}$$

where ν_i $(i=1,\ldots,s)$ are the partial multiplicities of $L(\lambda)$ at λ_0. In this case the set (1.4) of Jordan chains is called *canonical*. It is convenient to express the spectral data of λ_0 in terms of pairs of matrices $(\Phi(\lambda_0), J(\lambda_0))$, where

$$\Phi(\lambda_0) = [\phi_{10}\cdots\phi_{1,\nu_1-1}\phi_{20}\cdots\phi_{2,\nu_2-1}\cdots\phi_{s0}\cdots\phi_{s,\nu_s-1}],$$
$$J(\lambda_0) = J_1 \dot{+} J_2 \dot{+} \cdots \dot{+} J_s ,$$

and J_i stands for an elementary (upper) Jordan cell of size $\nu_i x \nu_i$ with eigenvalue λ_0. Any admissible pair that is similar to the pair $(\Phi(\lambda_0), J(\lambda_0)$ is called a *right Jordan pair of* $L(\lambda)$ *at* λ_0. Constructing analogously pairs $(\Phi(\lambda_i), J(\lambda_i))$ at every eigenvalue λ_i of $L(\lambda)$, one forms the matrices

$$\Phi_F = [\Phi(\lambda_1)\Phi(\lambda_2)\ldots\Phi(\lambda_p)], \quad J_F = \mathrm{diag}(J(\lambda_1), J(\lambda_2), \ldots, J(\lambda_p)),$$

where p is the number of different eigenvalues of $L(\lambda)$. The size of J_F coincides with the degree of the scalar polynomial $\det L(\lambda)$. Any admissible pair that is similar to (Φ_F, J_F) is called a *right finite Jordan pair of* $L(\lambda)$.

If $L(\lambda)$ is a monic matrix polynomial, then it is uniquely determined by its right finite Jordan pair. In this case a right finite Jordan pair is also called a *right standard pair* of the monic polynomial $L(\lambda)$. An important example of a right standard pair of the monic polynomial $L(\lambda) = \lambda^n I + \sum_{j=0}^{n-1} \lambda^j \ell_j$ is provided by the *companion standard pair* $(\Phi^{(0)}, C_L)$, *where*

$$\Phi^{(0)} = \underbrace{[I \ 0 \ldots 0]}_{n}, \quad C_L = \begin{bmatrix} 0 & I & & \\ & \ddots & \ddots & \\ & & & I \\ -\ell_0 & -\ell_1 & \cdots & -\ell_{n-1} \end{bmatrix}, \tag{1.5}$$

The matrix C_L will be called the *first companion matrix* of $L(\lambda)$. If $L(\lambda)$ is a matrix polynomial of degree n with invertible leading coefficient, its right finite Jordan pair coincides with the right standard pair of the monic polynomial $\hat{L}(\lambda) := \ell_n^{-1}L(\lambda)$, and hence in this case $L(\lambda)$ is determined by its right Jordan pair uniquely up to multiplication from the left by a scalar invertible matrix. This is not true any more if the leading coefficient of $L(\lambda)$ is not invertible. For a general regular polynomial $L(\lambda)$ of degree n one has to consider in addition to (Φ_F, J_F) also a *right Jordan pair at infinity* (Φ_∞, J_∞) which is defined as a right Jordan pair at $\lambda = 0$ of the polynomial $L^0(\lambda) = \lambda^n L(\lambda^{-1})$. If the point $\lambda = 0$ is not an eigenvalue of the polynomial $L(\lambda)$, then one defines a *right canonical pair* $(\Phi^{(L)}, J^{(L)})$ of $L(\lambda)$ by setting $\Phi^{(L)} = [\Phi_F\ \Phi_\infty]$ and $J^{(L)} = J_F^{-1} \dotplus J_\infty$. It is clear that an admissible pair (Φ, J) is a canonical pair of $L(\lambda)$ if and only if it is a standard pair of the monic polynomial $\hat{L}^0(\lambda) := \lambda^n \ell_0^{-1} L(\lambda^{-1})$.

Concerning the divisibility theory of matrix polynomials we need the following notions and results (see [11-12], [13] for details and proofs). Given regular matrix polynomials $L(\lambda)$ and $D(\lambda)$, we say that $D(\lambda)$ is a right divisor of $L(\lambda)$ if there is a matrix polynomial $Q(\lambda)$ such that $L(\lambda) = Q(\lambda)D(\lambda)$. A matrix polynomial $D(\lambda)$ is called a *common right divisor* of matrix polynomials $L_1(\lambda)$ and $L_2(\lambda)$ if $D(\lambda)$ is a right divisor of each $L_i(\lambda)$ $(i = 1,2)$. A right common divisor $D_0(\lambda)$ of the polynomials $L_1(\lambda)$, $L_2(\lambda)$ is called a *right greatest common divisor* of these polynomials if any other right common divisor of $L_1(\lambda), L_2(\lambda)$ is a right divisor of $D_0(\lambda)$ as well. If $D_0(\lambda) \equiv I$ is a greatest common right divisor of $L_1(\lambda)$ and $L_2(\lambda)$, we say that the polynomials $L_1(\lambda)$ and $L_2(\lambda)$ are *right coprime*. Next, a regular matrix polynomial $M(\lambda)$ is called a *common left multiple* of $L_1(\lambda), L_2(\lambda)$ if each $L_i(\lambda)$ $(i = 1,2)$ is a right divisor of $M(\lambda)$. If, in addition, $M(\lambda)$ is a right divisor of any (regular) common left multiple of $L_1(\lambda)$, $L_2(\lambda)$, then $M(\lambda)$ is said to be a *least common left multiple* of the polynomials $L_1(\lambda)$ and $L_2(\lambda)$.

In order to express the above notions in terms of Jordan pairs we need the following definitions. Given admissible pairs (Φ_1,Q_1) and (Φ_2,Q_2) of orders p_1 and p_2 ($p_1 \geqslant p_2$), respectively, we say that the pair (Φ_1,Q_1) is an *extension* of (Φ_2,Q_2) or, which is equivalent, (Φ_2,Q_2) is a *restriction* of (Φ_1,Q_1), if there exists a $p_1 x p_2$ matrix S of full rank such that $\Phi_1 S = \Phi_2$ and $Q_1 S = SQ_2$. In other words, the pair (Φ_1,Q_1) is an extension of (Φ_2,Q_2) if it is similar to a pair of the form

$$\left([\Phi_2 \ \Phi_2], \begin{bmatrix} Q_2 & \hat{Q}_2 \\ 0 & \tilde{Q}_2 \end{bmatrix} \right),$$

where $\Phi_2, \tilde{Q}_2, \hat{Q}_2$ are some matrices of appropriate sizes. A pair (Φ,Q) is called a *common restriction* of the admissible pairs (Φ_1,Q_1) and (Φ_2,Q_2) if each of these pairs is an extension of (Φ,Q). A common restriction of the pairs (Φ_1,Q_1) and (Φ_2,Q_2), which is an extension of any other common restriction of these pairs, is referred to as a *greatest common restriction* of the pairs (Φ_1,Q_1) and (Φ_2,Q_2). An admissible pair (Φ,Q) is said to be a *common extension* of (Φ_1,Q_1) and (Φ_2,Q_2) if (Φ,Q) is an extension of each (Φ_i,Q_i) ($i = 1,2$). We call (Φ_0,Q_0) a *least common extension* of (Φ_1,Q_1) and (Φ_2,Q_2) if any common extension of $(\Phi_1,Q_1),(\Phi_2,Q_2)$ is an extension of (Φ_0,Q_0) as well.

The following result is basic for the divisibility theory: a regular matrix polynomial $D(\lambda)$ is a right divisor of a regular matrix polynomial $L(\lambda)$ if and only if the right finite Jordan pair of $D(\lambda)$ is a restriction of the right finite Jordan pair of $L(\lambda)$. Then it follows that a regular polynomial $D(\lambda)$ is a (greatest) common right divisor of regular polynomials $L_1(\lambda)$ and $L_2(\lambda)$ if and only if the right finite Jordan pair of $D(\lambda)$ is a (greatest) common restriction of the right finite Jordan pairs of

$L_1(\lambda)$ and $L_2(\lambda)$. In particular, the polynomials $L_1(\lambda)$ and $L_2(\lambda)$ are right coprime if and only if their finite right Jordan pairs do not have any common restriction. Also, a regular polynomial $M(\lambda)$ is a (least) left common multiple of regular polynomials $L_1(\lambda)$, $L_2(\lambda)$ if and only if the right finite Jordan pair of $M(\lambda)$ is a (least) common extension of the right finite Jordan pairs of $L_1(\lambda)$ and $L_2(\lambda)$.

Note that the above "right" notions and results have appropriate "left" analogues. Recall that the row vectors $\phi_0, \phi_1, \ldots, \phi_k$ form a *left Jordan chain* of $L(\lambda)$ corresponding to λ_0 if and only if the column vectors $\phi_0^T, \phi_1^T, \ldots, \phi_k^T$ form a right Jordan chain of $L^T(\lambda) := \sum_{j=0}^{n} \lambda^j \ell_j^T$ corresponding to the same λ_0. Using left Jordan chains, one introduces in an obvious way the notions of left Jordan pairs, left standard pairs, etc. We mention only that the left admissible pair $(\hat{C}_L, \Psi^{(0)})$, where

$$\hat{C}_L = \begin{bmatrix} 0 & & & -\ell_0 \\ I & \ddots & & -\ell_1 \\ & \ddots & \ddots & \vdots \\ & & I & -\ell_{n-1} \end{bmatrix}, \qquad \Psi^{(0)} = \begin{bmatrix} I \\ 0 \\ \vdots \\ 0 \end{bmatrix} \tag{1.6}$$

provides an example of a *left standard pair* of the monic polynomial $L(\lambda) = \lambda^n I + \sum_{j=0}^{n-1} \lambda^j \ell_j$. The matrix $\hat{C}_L$ will be referred to as the *second companion matrix* of $L(\lambda)$.

2. THE FIRST M.G.KREIN THEOREM FOR BLOCK TOEPLITZ MATRICES

In this section we prove the following matrix generalization of Theorem 0.1.

THEOREM 2.1. *Let* $T = T_n = [t_{j-k}]_{j,k=0}^{n}$ *be an* $(n+1)r\times(n+1)r$ *hermitian block Toeplitz matrix, and let* $x = \mathrm{col}(x_j)_{j=0}^{n}$ *be a solution of the equation*

$$T_n \mathrm{col}(x_j)_{j=0}^{n} = \mathrm{col}(\delta_{0j} I)_{j=0}^{n}. \tag{2.1}$$

If x_0 *is positive (negative) definite, then the matrix polynomial*

$X^0(\lambda) = \sum_{j=0}^{n} \lambda^j x_{n-j}$ *does not have eigenvalues on the unit circle and the number of zeroes (counting multiplicities) of* $\det X^0(\lambda)$ *inside the unit circle is equal to the number of positive (negative) eigenvalues of the matrix* $T_{n-1} = [t_{j-k}]_{j,k=0}^{n-1}$.

In order to check that this theorem is indeed a generalization of Theorem 0.1, we note that if t_j $(j = 0,\pm1,\ldots\pm n)$ are complex numbers and $D_k = \det T_k \neq 0$ $(k=0,1,\ldots,n)$, then $x_0 = \frac{D_{n-1}}{D_n}$ and, by G. Jacobi's theorem, the number of positive (negative) eigenvalues of T_{n-1} is equal to the number of constancies (alterations) of sign in the sequence $1,D_1,\ldots,D_{n-1}$ (see e.g., [22], p.296]). The next example (which we borrow from [1]) shows that if x_0 is not definite, then $X^0(\lambda)$ may even have eigenvalues on the unit circle. Indeed, take

$$t_0 = \begin{bmatrix} 1 & 0 \\ 0 & -1 \end{bmatrix}, \qquad t_1 = \begin{bmatrix} 0 & 1 \\ 1 & 0 \end{bmatrix}.$$

Then

$$\begin{bmatrix} t_0 & t_1 \\ t_1 & t_0 \end{bmatrix} \begin{bmatrix} \frac{1}{2}t_0 \\ \frac{1}{2}t_1 \end{bmatrix} = \begin{bmatrix} I \\ 0 \end{bmatrix},$$

and hence the polynomial $X^0(\lambda) = \frac{1}{2}\begin{bmatrix} \lambda & 1 \\ 1 & -\lambda \end{bmatrix}$ has eigenvalues $\lambda = \pm i$.

Now introduce some useful notations. For a monic $r\times r$ matrix polynomial $A(\lambda) = \lambda^n I + \sum_{j=0}^{n-1} \lambda^j a_j$ we introduce two additional companion matrices:

$$\hat{K}_A := \begin{bmatrix} -a_{n-1} & -a_{n-2} & \cdots & -a_0 \\ I & & & 0 \\ & \ddots & & \vdots \\ & & I & 0 \end{bmatrix}, \quad K_A := \begin{bmatrix} -a_{n-1} & I & & \\ -a_{n-2} & & \ddots & \\ \vdots & & & I \\ -a_0 & 0 & \cdots & 0 \end{bmatrix}. \tag{2.2}$$

The relationship between the above matrices and the usual companion matrices C_A and $\hat{C}_A$ (see (1.5),(1.6)) is given by the equalities

$$P\hat{K}_A P = C_A , \quad PK_A P = \hat{C}_A , \tag{2.3}$$

where P denotes the reverse identity block matrix:

$$P = \begin{bmatrix} & & I \\ & \cdot^{\cdot^{\cdot}} & \\ I & & \end{bmatrix} .$$

We shall also need the following relation between the first and the second companion matrices (see e.g., [21]):

$$C_A = S_A^{-1} \hat{C}_A S_A , \tag{2.4}$$

where

$$S_A := \begin{bmatrix} a_1 & a_2 & \cdots & a_{n-1} & I \\ a_2 & & \cdot^{\cdot^{\cdot}} & \cdot^{\cdot^{\cdot}} & \\ \vdots & \cdot^{\cdot^{\cdot}} & \cdot^{\cdot^{\cdot}} & & \\ a_{n-1} & \cdot^{\cdot^{\cdot}} & & & \\ I & & & & \end{bmatrix} .$$

If $B(\lambda) = \sum_{j=0}^{n} \lambda^j b_j$ is a matrix polynomial of degree n with invertible leading coefficient b_n, we set

$$\hat{B}(\lambda) = b_n^{-1} B(\lambda), \quad \tilde{B}(\lambda) = B(\lambda) b_n^{-1} . \tag{2.5}$$

PROOF OF THEOREM 2.1. The proof of the theorem is divided into three steps.

Step 1. In this step we prove that if $T_n = (t_{j-k})_{j,k=0}^{n}$ is a block-Toeplitz matrix (not necessarily hermitian), and if $x = \mathrm{col}(x_j)_{j=0}^{n}$ is a solution of the equation (2.1) and $v = \mathrm{row}(v_j)_{j=0}^{n}$ is a solution of the equation

$$[v_0 v_1 \ldots v_n] T = [I\ 0\ \ldots\ 0] , \tag{2.6}$$

such that the matrix $x_0 = v_0$ is invertible, then the matrix $T_{n-1} = (t_{j-k})_{j,k=0}^{n-1}$ satisfies the following Stein type equation (cf.[21]):

$$T_{n-1} - \hat{K}_{\hat{V}^0} T_{n-1} K_{\tilde{X}^0} = \operatorname{diag}(v_0^{-1}, 0, \dots, 0). \tag{2.7}$$

Indeed, equations (2.1) and (2.6) can be rewritten in the form

$$t_0 x_0 + \operatorname{row}(t_{-j})_{j=1}^n \cdot \operatorname{col}(x_j)_{j=1}^n = I, \tag{2.8}$$

$$\operatorname{col}(t_j)_{j=1}^n = -T_{n-1} \operatorname{col}(x_j x_0^{-1})_{j=1}^n, \tag{2.9}$$

and

$$v_0 t_0 + \operatorname{row}(v_j)_{j=1}^n \cdot \operatorname{col}(t_j)_{j=1}^n = I, \tag{2.10}$$

$$\operatorname{row}(t_{-j})_{j=1}^n = -\operatorname{row}(v_0^{-1} v_j)_{j=1}^n T_{n-1}, \tag{2.11}$$

respectively. Substituting (2.11) in (2.8) we obtain

$$t_0 - \operatorname{row}(v_0^{-1} v_j)_{j=1}^n T_{n-1} \operatorname{col}(x_j x_0^{-1}) = v_0^{-1}. \tag{2.12}$$

A simple computation shows that

$$\hat{K}_{\hat{V}^0} T_{n-1} K_{\tilde{X}^0} = \begin{bmatrix} \operatorname{row}(v_0^{-1} v_j)_{j=1}^n T_{n-1} \operatorname{col}(x_j x_0^{-1})_{j=1}^n & -\operatorname{row}(v_0^{-1} v_j)_{j=1}^n T_{n-1} \Delta^\tau \\ -\Delta T_{n-1} \operatorname{col}(x_j x_0^{-1})_{j=1}^n & \Delta T_{n-1} \Delta^\tau \end{bmatrix} \tag{2.13}$$

where

$$\Delta := \underbrace{\begin{bmatrix} I & 0 & \cdots & & 0 \\ & I & & & 0 \\ & & \ddots & & \vdots \\ & & & I & 0 \end{bmatrix}}_{n} \Big\} n-1 \, .$$

Note that $\Delta T_{n-1} \Delta^\tau = T_{n-2}$. Thus, substituting (2.9), (2.11) and (2.12) in (2.13), we obtain

$$\hat{K}_{\hat{V}^0} T_{n-1} K_{\tilde{X}^0} = \left[\begin{array}{c|ccc} t_0 - v_0^{-1} & t_{-1} & \cdots & t_{-n} \\ \hline t_1 & & & \\ \vdots & & T_{n-2} & \\ t_{n-1} & & & \end{array} \right],$$

which is just the same as (2.7).

Step 2. In this step we prove that by preserving the notations and assumptions of Step 1, the matrix T_{n-1} satisfies the equation

$$T_{n-1} - \hat{K}^{n}_{\hat{V}0} T_{n-1} K^{n}_{\tilde{X}0} =$$

$$\begin{bmatrix} I & v_0^{-1}v_1 & \cdots & v_0^{-1}v_{n-1} \\ & \ddots & \ddots & \vdots \\ & & \ddots & v_0^{-1}v_1 \\ & & & I \end{bmatrix}^{-1} \begin{bmatrix} x_0^{-1} & & & \\ & x_0^{-1} & & \\ & & \ddots & \\ & & & x_0^{-1} \end{bmatrix} \begin{bmatrix} I & & & \\ x_1x_0^{-1} & \ddots & & \\ \vdots & \ddots & \ddots & \\ x_{n-1}x_0^{-1} & \cdots & x_1x_0^{-1} & I \end{bmatrix}^{-1} \tag{2.14}$$

Indeed, denoting the right hand side of (2.7) by R, and multiplying both sides of (2.7) by $\hat{K}_{\hat{V}0}$ and $K_{\tilde{X}0}$, from the left and from the right, respectively, we obtain

$$\hat{K}_{\hat{V}0}T_{n-1}K_{\tilde{X}0} = \hat{K}^{2}_{\hat{V}0}T_{n-1}K^{2}_{\tilde{X}0} + \hat{K}_{\hat{V}0}RK_{\tilde{X}0} ,$$

and in view of (2.7) we infer that

$$T_{n-1} - \hat{K}^{2}_{\hat{V}0}T_{n-1}K^{2}_{\tilde{X}0} = R + \hat{K}_{\hat{V}0}RK_{\tilde{X}0} .$$

Proceeding in this way we obtain

$$T_{n-1} - \hat{K}^{n}_{\hat{V}0}T_{n-1}K^{n}_{\tilde{X}0} = \sum_{i=o}^{n-1} \hat{K}^{i}_{\hat{V}0}RK^{i}_{\tilde{X}0}. \tag{2.15}$$

Now denote

$$\Psi^{(i)} = \operatorname{col}(\delta_{ik}I)_{k=0}^{n-1} , \quad \Phi^{(i)} = \operatorname{row}(\delta_{ik}I)_{k=0}^{n-1} \quad (i=0,1,\dots,n-1)$$

and note that $R = \Psi^{(0)}x_0^{-1}\Phi^{(0)}$. Using (2.3) and (2.4) we see that

$$\hat{K}^{i}_{\hat{V}0}\Psi^{(0)} = PC_{\hat{V}0}P\Psi^{(0)} = PS^{-1}_{\hat{V}0}\hat{C}^{i}_{\hat{V}0}S_{\hat{V}0}P\Psi^{(0)} = PS^{-1}_{\hat{V}0}\Psi^{(i)} \quad (i = 0,1,\dots,n-1).$$

Similarly,

$$\Phi^{(0)}K^{i}_{\tilde{X}0} = \Phi^{(i)}S^{-1}_{\tilde{X}0}P \quad (i = 0,1,\dots,n-1).$$

Thus the right hand side in (2.15) can be written in the form

$$PS^{-1}_{\hat{V}0}\Big(\sum_{i=o}^{n-1} \Psi^{(i)}x_0^{-1}\Phi^{(i)}\Big)S^{-1}_{\tilde{X}0}P,$$

which clearly coincides with the expression on the right in (2.14).

Step 3. In this step we complete the proof of Theorem 2.1 using the following well known inertia theorem (see [18],[19],[26],[28],[29]).

If A is a pxp complex matrix such that $H - A^*HA > 0$ for some hermitian matrix H, then H is nonsingular and the number of eigenvalues of A inside (outside) the unit circle is equal to the number of positive (negative) eigenvalues of H.

Now let us write the equation (2.14) in the case of a hermitian matrix T. In this case the solutions of (2.1) and (2.6) are related by the obvious relation $v_j = x_j^*$ and hence (2.14) becomes

$$T_{n-1} - (K_{\tilde{X}^0}^*)^n T_{n-1} K_{\tilde{X}^0}^n = \Gamma^* \mathrm{diag}(x_0^{-1}, \dots, x_0^{-1}) \Gamma , \qquad (2.16)$$

where

$$\Gamma = \begin{bmatrix} I & & & \\ \ddots & \ddots & & \\ x_1 x_0^{-1} & \ddots & \ddots & \\ \vdots & \ddots & \ddots & \\ x_n x_0^{-1} & \cdots & x_1 x_0^{-1} & I \end{bmatrix}^{-1} . \qquad (2.17)$$

Since $x_0 > 0$ we infer from the inertia theorem cited above that the matrix $K_{\tilde{X}^0}^n$ has no eigenvalues on the unit circle and the number of its eigenvalues inside the unit circle is equal to the number of positive eigenvalues of T_{n-1}. It is clear that the same assertions hold true for the matrix $K_{\tilde{X}^0}$ itself and, in view of (2.3), for the companion matrix $\hat{C}_{\tilde{X}^0}$. Since $\det X^0(\lambda)$ and $\det(\hat{C}_{\tilde{X}^0} - \lambda I)$ coincide up to a constant factor, the theorem is proved.

□

3. THE BASIC INVERSE PROBLEM FOR INVERTIBLE BLOCK-TOEPLITZ MATRICES

The inverse problems which are considered in this section deal with constructing an invertible block Toeplitz matrix T via a relatively small number of block columns and block rows of its inverse T^{-1}. The basic problem of this type (in the block matrix case) is the one in which 4 block-vectors are given: the first and the last block rows and block columns of T^{-1}. In this section we state the Gohberg-Heinig theorem [10], which solves the above problem, and we present a modification of this theorem, which is useful for our purposes. Finally, we specify this modification for hermitian block Toeplitz matrices.

In what follows we deal with matrices whose entries are rxr complex matrices themselves. The following result, due to I. Gohberg and G. Heinig [10], is basic for us.

THEOREM 3.1. *Let* $x = \mathrm{col}(x_j)_{j=0}^{n}$, $y = \mathrm{col}(y_j)_{j=0}^{n}$, $u = \mathrm{row}(u_j)_{j=0}^{n}$ *and* $v = \mathrm{row}(v_j)_{j=0}^{n}$ *be given block vectors such that* x_0 *and* u_n *are invertible. There exists a block Toeplitz matrix* $T = [t_{p-q}]_{p,q=0}^{n}$ *satisfying*

$$T\begin{bmatrix} x_0 \\ x_1 \\ \vdots \\ x_n \end{bmatrix} = \begin{bmatrix} I \\ 0 \\ \vdots \\ 0 \end{bmatrix}, \qquad T\begin{bmatrix} y_0 \\ \vdots \\ y_{n-1} \\ y_n \end{bmatrix} = \begin{bmatrix} 0 \\ \vdots \\ 0 \\ I \end{bmatrix}, \tag{3.1}$$

$$[u_0 \dots u_{n-1} u_n]T = [0 \dots 0\ I], \qquad [v_0 v_1 \dots v_n]T = [\ I\ 0 \dots 0\ ] \tag{3.2}$$

if and only if the following conditions hold:

(a) $v_0 = x_0$, $y_n = u_n$;

(b)
$$\begin{bmatrix} y_0 & & 0 \\ \vdots & \ddots & \\ y_n & \cdots & y_0 \end{bmatrix} u_n^{-1} \begin{bmatrix} u_n & & 0 \\ \vdots & \ddots & \\ u_0 & \cdots & u_n \end{bmatrix} = \begin{bmatrix} x_0 & & 0 \\ \vdots & \ddots & \\ x_n & \cdots & x_0 \end{bmatrix} x_0^{-1} \begin{bmatrix} v_n & & 0 \\ \vdots & \ddots & \\ v_0 & \cdots & v_n \end{bmatrix},$$

$$\begin{bmatrix} y_n & & 0 \\ \vdots & \ddots & \\ y_0 & \cdots & y_n \end{bmatrix} u_n^{-1} \begin{bmatrix} u_0 & & 0 \\ \vdots & \ddots & \\ u_n & \cdots & u_0 \end{bmatrix} = \begin{bmatrix} x_n & & 0 \\ \vdots & \ddots & \\ x_0 & \cdots & x_n \end{bmatrix} x_0^{-1} \begin{bmatrix} v_0 & & 0 \\ \vdots & \ddots & \\ v_n & \cdots & v_0 \end{bmatrix};$$

(c) at least one of the following matrices is invertible:

$$M = \underbrace{\begin{bmatrix} x_0 & & & y_0 & & \\ \vdots & \ddots & & \vdots & \ddots & \\ & & x_0 & & & y_0 \\ x_n & & \vdots & y_n & & \vdots \\ & \ddots & & & \ddots & \\ & & x_n & & & y_n \end{bmatrix}}_{2n}, \qquad N = \left.\begin{bmatrix} u_n & \cdots & u_0 & & & \\ & \ddots & & \ddots & & \\ & & u_n & \cdots & & u_0 \\ v_n & \cdots & v_0 & & & \\ & \ddots & & \ddots & & \\ & & v_n & \cdots & & v_0 \end{bmatrix}\right\} 2n .$$

If the conditions (a)-(c) are fulfilled, then T *is invertible and can be obtained as the inverse of the following matrix:*

$$T^{-1} = \begin{bmatrix} x_0 & & & 0 \\ x_1 & \ddots & & \\ \vdots & \ddots & \ddots & \\ x_n & \cdots & x_1 & x_0 \end{bmatrix} v_0^{-1} \begin{bmatrix} v_0 & v_1 & \cdots & v_n \\ & \ddots & \ddots & \vdots \\ & & \ddots & v_1 \\ 0 & & & v_0 \end{bmatrix} - \begin{bmatrix} 0 & & & 0 \\ y_0 & \ddots & & \\ \vdots & \ddots & \ddots & \\ y_{n-1} & \cdots & y_0 & 0 \end{bmatrix} y_n^{-1} \begin{bmatrix} 0 & u_0 & \cdots & u_{n-1} \\ & \ddots & \ddots & \vdots \\ & & \ddots & u_0 \\ 0 & & & 0 \end{bmatrix} \tag{3.3}$$

It is proved in [10] that if the conditions of Theorem 3.1 are fulfilled, then the matrix $T_{n-1} = [T_{j-k}]_{j,k=0}^{n-1}$ is also invertible and one can find its inverse by the following formula:

$$T_{n-1}^{-1} = \begin{bmatrix} x_0 & & 0 \\ \vdots & \ddots & \\ x_{n-1} & \cdots & x_0 \end{bmatrix} v_0^{-1} \begin{bmatrix} v_0 & \cdots & v_{n-1} \\ & \ddots & \vdots \\ 0 & & v_0 \end{bmatrix} - \begin{bmatrix} y_0 & & 0 \\ \vdots & \ddots & \\ y_{n-1} & \cdots & y_0 \end{bmatrix} y_n^{-1} \begin{bmatrix} u_0 & \cdots & u_{n-1} \\ & \ddots & \vdots \\ 0 & & u_0 \end{bmatrix} \tag{3.3'}$$

Now our aim is to express conditions (b) and (c) of the preceding theorem in terms of the matrix polynomials

$$X(\lambda) := \sum_{j=0}^{n} \lambda^j x_j, \qquad Y(\lambda) := \sum_{j=0}^{n} \lambda^j y_j$$
$$U^O(\lambda) := \sum_{j=0}^{n} \lambda^j u_{n-j}, \qquad V^O(\lambda) := \sum_{j=0}^{n} \lambda^j v_{n-j}. \tag{3.4}$$

A simple verification shows that condition (b) can be rewritten in the form (b') $$Y(\lambda)u_n^{-1}U^0(\lambda) = X(\lambda)x_0^{-1}V^0(\lambda).$$

To express condition (c) in terms of the polynomials involved we need some additional facts about resultant matrices. For two r×r matrix polynomials $L(\lambda) = \sum_{j=o}^{\nu} \lambda^j \ell_j$ and $M(\lambda) = \sum_{j=o}^{\mu} \lambda^j m_j$ the corresponding *resultant matrices* $R_q(L,M)$ are defined by

$$R_q(L,M) := \underbrace{\begin{bmatrix} \ell_0 & \ell_1 & \cdots & \ell_\nu & & & \\ & \ell_0 & \ell_1 & & \ell_\nu & & \\ & & \ddots & \ddots & & \ddots & \\ & & & \ell_0 & \ell_1 & \cdots & \ell_\nu \\ m_0 & m_1 & \cdots & m_\mu & & & \\ & m_0 & m_1 & \cdots & m_\mu & & \\ & & \ddots & \ddots & & \ddots & \\ & & & m_0 & m_1 & \cdots & m_\mu \end{bmatrix}}_{q \text{ blocks}},$$

where $q > \max\{v,\mu\}$. The following property of the resultant matrices is basic (see [11], [12] for details):

Let (Φ_F, J_F) (resp., (Φ_∞, J_∞)) denote the greatest common restriction of right finite (resp. infinite) Jordan pairs of the (regular) matrix polynomials $L(\lambda)$ and $M(\lambda)$. Then there is an integer $q_0(L,M)$ such that $\dim\operatorname{Ker}R_q(L,M) < \dim\operatorname{Ker}R_{q-1}(L,M)$ for any $q < q_0(L,M)$, while for every $q \geqslant q_0(L,M)$ the following representation holds true:

$$\operatorname{Ker}R_q(L,M) = \operatorname{Im}\operatorname{col}\left(\Phi_F J_F^{\,i-1}\right)_{i=1}^q \dotplus \operatorname{Im}\operatorname{col}\left(\Phi_\infty J_\infty^{\,q-i}\right)_{i=1}^q. \tag{3.5}$$

The integer $q_0(L,M)$ can be explicitly expressed via the spectral data of the polynomials $L(\lambda)$ and $M(\lambda)$. For the sake of simplicity assume that the point $\lambda = 0$ is not an eigenvalue of $L(\lambda)$ and $M(\lambda)$, and let $(\Phi^{(L)}, J^{(L)})$ and $(\Phi^{(M)}, J^{(M)})$ be the canonical pairs of $L(\lambda)$ and $M(\lambda)$, respectively. Put

$$\Phi = [\Phi^{(L)}\Phi^{(M)}], \quad J = J^{(L)} \dotplus J^{(M)}. \tag{3.6}$$

Then

$$q_0(L,M) = \max\{\mathrm{ind}(\Phi,J),\ \nu+1,\mu+1\}. \tag{3.7}$$

It turns out that if the polynomials $L(\lambda)$ and $M(\lambda)$ satisfy an equation of type (b'), one can obtain a simple estimation for the integer $q_0(L,M)$. This follows from the next result.

PROPOSITION 3.2. *Let* $L(\lambda = \sum_{j=0}^{\nu} \lambda^j \ell_j$ *and* $M(\lambda) = \sum_{j=0}^{\mu} \lambda^j m_j$ *be regular matrix polynomials and assume that*

$$M_1(\lambda)L(\lambda) = L_1(\lambda)M(\lambda) \tag{3.8}$$

for some regular matrix polynomials $M_1(\lambda) = \sum_{j=0}^{\mu_1} \lambda^j m_j^{(1)}$, $L_1(\lambda = \sum_{j=0}^{\nu_1} \lambda^j \ell_j^{(1)}$.

Then

$$q_0(L,M) \leqslant \max\{\nu+\mu_1,\mu+\nu_1,\nu+1,\mu+1\}. \tag{3.9}$$

In particular, if $\mu_1 = \mu$, $\nu_1 = \nu$ *and* $\mu+\nu \geqslant 1$, *then (3.5) holds true for the square matrix* $R_{\nu+\mu}(L,M)$.

PROOF. We shall prove the proposition for the case of comonic polynomials $L(\lambda)$ and $M(\lambda)$. The general case is reduced to the comonic one by a standard argument of replacing regular polynomials $L(\lambda)$, $M(\lambda)$ by the comonic polynomials $L^{-1}(\alpha)L(\lambda+\alpha)$ and $M^{-1}(\alpha)M(\lambda+\alpha)$, where $\alpha \notin \sigma(L) \cup \sigma(M)$ (see [11], [12], [13]).

So let $L(\lambda)$ and $M(\lambda)$ be comonic polynomials. Denote $\gamma = \max\{\nu+\mu_1,\ \mu+\nu_1\}$ and consider $L_1(\lambda)$ and $M_1(\lambda)$ as polynomials of degree $\gamma-\mu$ and $\gamma-\nu$, respectively, so that $L_1^O(\lambda) = \lambda^{\gamma-\mu}L_1(\lambda^{-1})$ and $M_1^O(\lambda) = \lambda^{\gamma-\nu}M_1(\lambda^{-1})$. Then rewrite equation (3.8) as

$$M_1^O(\lambda)L^O(\lambda) = L_1^O(\lambda)M^O(\lambda)\ \big(=: \Gamma(\lambda)\big). \tag{3.10}$$

It follows that the (monic) polynomials $L^O(\lambda)$ and $M^O(\lambda)$ have a left common multiple $\Gamma(\lambda)$ of degree γ. If $\big(\Phi^{(\Gamma)},J^{(\Gamma)}\big)$ is the finite Jordan pair of $\Gamma(\lambda)$, then its index of stabilization is less than or equal to the degree of $\Gamma(\lambda)$ (see [11], [12], [13]):

$$\gamma \geqslant \mathrm{ind}\left(\Phi^{(\Gamma)}, J^{(\Gamma)}\right). \tag{3.11}$$

Next, if $\tilde{\Gamma}(\lambda)$ denotes a least left common multiple of $L^{O}(\lambda)$ and $M^{O}(\lambda)$, and if $\left(\Phi^{(\tilde{\Gamma})}, J^{(\tilde{\Gamma})}\right)$ stands for a finite Jordan pair of $\tilde{\Gamma}(\lambda)$, then $\left(\Phi^{(\tilde{\Gamma})}, J^{(\tilde{\Gamma})}\right)$ is a restriction of $\left(\Phi^{(\Gamma)}, J^{(\Gamma)}\right)$, and, clearly

$$\mathrm{ind}\left(\Phi^{(\Gamma)}, J^{(\Gamma)}\right) \geqslant \mathrm{ind}\left(\Phi^{(\tilde{\Gamma})}, J^{(\tilde{\Gamma})}\right). \tag{3.12}$$

The pair $\left(\Phi^{(\tilde{\Gamma})}, J^{(\tilde{\Gamma})}\right)$ is a least common extension of the standard pairs $\left(\Phi^{(L)}, J^{(L)}\right)$ and $\left(\Phi^{(M)}, J^{(M)}\right)$ of the polynomials $L^{O}(\lambda)$ and $M^{O}(\lambda)$, respectively. Hence Theorem 2.3 from [11] implies that

$$\mathrm{ind}\left(\Phi^{(\tilde{\Gamma})}, J^{(\tilde{\Gamma})}\right) = \mathrm{ind}(\Phi, J), \tag{3.13}$$

where the pair (Φ, J) is defined by (3.6). Combining (3.11)-(3.13) we see that $\gamma \geqslant \mathrm{ind}(\Phi, J)$, and in view of (3.7) we have

$$q_0(L,M) \leqslant \max\{\gamma, \nu+1, \mu+1\},$$

which proves the Proposition.

□

Now we are in a position to state a modification of Theorem 3.1 in terms of the matrix polynomials defined by (3.4).

THEOREM 3.3. *Let* x, y, u *and* v *be as in Theorem 3.1, and let the matrix polynomials* $X(\lambda)$, $Y(\lambda)$, $U^{O}(\lambda)$ *and* $V^{O}(\lambda)$ *be defined by (3.4). There exists a block Toeplitz matrix* T *satisfying (3.1)-(3.2) if and only if the following conditions are fulfilled:*

(a') $v_0 = x_0$, $y_n = u_n$;

(b') $Y(\lambda) y_n^{-1} U^{O}(\lambda) = X(\lambda) x_0^{-1} V^{O}(\lambda)$;

(c') $Y(\lambda)$ *and* $X(\lambda)$ *are left coprime or (which is the same)* $U^{O}(\lambda)$ *and* $V^{O}(\lambda)$ *are right coprime.*

If conditions (a')-(c') are fulfilled, then T *is invertible and can be found by virtue of (3.3).*

PROOF. As already mentioned, condition (b') is equivalent to condition (b) in Theorem 3.1. We shall show now that condition (c') is equivalent to condition (c) in Theorem 3.1, provided condition (b') is fulfilled. Indeed, the matrix N in (c) coincides with the resultant matrix $R_{2n}(U^0,V^0)$ of the polynomials $U^0(\lambda)$ and $V^0(\lambda)$. Since these polynomials satisfy equation (b'), we can apply Proposition 3.2 to conclude that

$$\operatorname{Ker} N = \operatorname{Im} \operatorname{col}\left(\Phi_F J_F^{i-1}\right)_{i=1}^{2n} \dotplus \operatorname{Im} \operatorname{col}\left(\Phi_\infty J_\infty^{2n-i}\right)_{i=1}^{2n}, \tag{3.14}$$

where (Φ_F, J_F) (resp., (Φ_∞, J_∞)) denotes the greatest common restriction of the finite (resp., infinite) right Jordan pairs of $U^0(\lambda)$ and $V^0(\lambda)$. Since the polynomials $U^0(\lambda)$ and $V^0(\lambda)$ have invertible leading coefficients, the point $\lambda = \infty$ is not an eigenvalue of these polynomials, and hence (3.14) becomes

$$\operatorname{Ker} N = \operatorname{Im} \operatorname{col}\left(\Phi_F J_F^{i-1}\right)_{i=1}^{2n}.$$

But the matrix N is square, and therefore the last equality shows that N is invertible if and only if the polynomials $U^0(\lambda)$ and $V^0(\lambda)$ are right coprime.

In a similar way one checks that the invertibility of the matrix M in condition (c) of Theorem 3.1 is equivalent to the left coprimeness of the polynomials $X(\lambda)$ and $Y(\lambda)$. Note also that the results of [23] show that the right coprimeness of $U^0(\lambda)$ and $V^0(\lambda)$ is equivalent to the left coprimeness of $X(\lambda)$ and $Y(\lambda)$, provided (b') is fulfilled. Now applying Theorem 3.1 one easily completes the proof of the theorem. □

Passing to the inverse problem in the case of a hermitian matrix T, we first observe that in this case the solutions of (3.2) can be obtained from the solutions of (3.1) by means of the relations

$$u_j = y_j^*, \quad v_j = x_j^* \quad (j = 0,1,\ldots,n).$$

So, Theorem 3.3 readily implies the following solution of the basic inverse problem for hermitian invertible block Toeplitz matrices.

THEOREM 3.4. *Let* $x = \operatorname{col}(x_j)_{j=0}^{n}$ *and* $y = \operatorname{col}(y_j)_{j=0}^{n}$ *be given block column vectors such that* x_0 *and* y_n *are invertible. There exists a hermitian block Toeplitz matrix* $T = [t_{j-k}]_{j,k=0}^{n}$ *such that*

$$T\begin{bmatrix} x_0 \\ x_1 \\ \vdots \\ x_n \end{bmatrix} = \begin{bmatrix} I \\ 0 \\ \vdots \\ 0 \end{bmatrix}, \quad T\begin{bmatrix} y_0 \\ \vdots \\ y_{n-1} \\ y_n \end{bmatrix} = \begin{bmatrix} 0 \\ \vdots \\ 0 \\ I \end{bmatrix}, \qquad (3.15)$$

if and only if the following conditions are satisfied:

(*i*) $x_0^* = x_0, \quad y_n^* = y_n$;

(*ii*) $Y(\lambda)y_n^{-1}Y^{\circledast}(\lambda) = X(\lambda)x_0^{-1}X^{\circledast}(\lambda)$;

(*iii*) $Y(\lambda)$ *and* $X(\lambda)$ *are left coprime or (which is equivalent)* $Y^{\circledast}(\lambda)$ *and* $X^{\circledast}(\lambda)$ *are right coprime.*

If conditions (i)-(iii) are fulfilled, then T *is invertible and can be obtained as the inverse of the following matrix*

$$T^{-1} = \begin{bmatrix} x_0 & & & 0 \\ x_1 & \ddots & & \\ \vdots & \ddots & \ddots & \\ x_n & \cdots & x_1 & x_0 \end{bmatrix} x_0^{-1} \begin{bmatrix} x_0^* & x_1^* & \cdots & x_n^* \\ & \ddots & \ddots & \vdots \\ & & \ddots & x_1^* \\ 0 & & & x_0^* \end{bmatrix} - \begin{bmatrix} 0 & & & 0 \\ y_0 & \ddots & & \\ \vdots & \ddots & \ddots & \\ y_{n-1} & \cdots & y_0 & 0 \end{bmatrix} y_n^{-1} \begin{bmatrix} 0 & y_0^* & \cdots & y_{n-1}^* \\ & \ddots & \ddots & \vdots \\ & & \ddots & y_0^* \\ 0 & & & 0 \end{bmatrix}. \qquad (3.16)$$

In the above statements we use for brevity the notation

$$X^{\circledast}(\lambda) := (X^{O})^*(\lambda) := \sum_{j=0}^{n} \lambda^j x_{n-j}^* .$$

Note that in the hermitian case, formula (3.3') for $T_{n-1} = [t_{j-k}]_{j,k=0}^{n-1}$ becomes

$$T_{n-1}^{-1} = \begin{bmatrix} x_o & & 0 \\ \vdots & \ddots & \\ x_{n-1} & \cdots & x_0 \end{bmatrix} v_0^{-1} \begin{bmatrix} x_0^* & \cdots & x_{n-1}^* \\ & \ddots & \vdots \\ 0 & & x_0^* \end{bmatrix} - \begin{bmatrix} y_o & & 0 \\ \vdots & \ddots & \\ y_{n-1} & \cdots & y_0 \end{bmatrix} y_n^{-1} \begin{bmatrix} y_0^* & \cdots & y_{n-1}^* \\ & \ddots & \vdots \\ 0 & & y_0^* \end{bmatrix}. \quad (3.16')$$

As a result of the last theorem, one can solve the inverse problem of determining an invertible hermitian block-Toeplitz matrix T via the first block column $\mathrm{col}(x_j)_{j=0}^n$ and the last block column $\mathrm{col}(y_j)_{j=0}^n$ of its inverse T^{-1}, provided these columns satisfy conditions (i)-(iii). In this case the matrix T is determined uniquely by the formula (3.16). In the more difficult case, when only the first column $\mathrm{col}(x_j)_{j=0}^n$ $(x_0^* = x_0,\ \det x_0 \neq 0)$ of T^{-1} (i.e., the solution of the first equation in (3.15)) is given, Theorem 3.4 can be used along the following lines. For the matrix polynomial $R(\lambda) := X(\lambda)x_0^{-1}X^{\circledast}(\lambda)$ $\left(X(\lambda) := \sum_{j=1}^{n} \lambda^j x_j\right)$ find a factorization of the form

$$R(\lambda) = Y(\lambda)y_n^{-1}Y^{\circledast}(\lambda),$$

where $Y(\lambda) = \sum_{j=0}^{n} \lambda^j y_j$ is a matrix polynomial with hermitian invertible leading coefficient y_n such that $X(\lambda)$ and $Y(\lambda)$ are left coprime, and then apply formula (3.16). In this way the problem of generalizing the second M.G. Krein theorem is reduced to the above factorization problem. This approach will be carried out in Section 6 of the present paper. To solve the above mentioned factorization problem we need some auxiliary results on solvability of equations with matrix polynomials and matrix equation, which is the topic of the next section.

4. EQUATIONS IN MATRIX POLYNOMIALS AND LINEAR MATRIX EQUATIONS

In this section we deal with the equation

$$M(\lambda)Y(\lambda) + Z(\lambda)L(\lambda) = R, \quad (E_1)$$

where the rxr matrix polynomials

$$M(\lambda) = \sum_{j=0}^{\mu} \lambda^j m_j, \qquad L(\lambda) = \sum_{j=0}^{\nu} \lambda^j \ell_j, \tag{4.1}$$

and the right-hand side $R \in \mathbb{C}^{r\times r}$ are given, while the rxr matrix polynomials $Y(\lambda)$ and $Z(\lambda)$ have to be found. If $Y(\lambda)$ and $Z(\lambda)$ satisfy (E_1) and $\deg Y \leq \nu-1$, $\deg Z \leq \mu-1$, we say that $(Y(\lambda),Z(\lambda))$ is a *$(\nu-1,\mu-1)$-solution* of (E_1).

We remark that if m_μ is invertible, equation (E_1) is solvable if and only if there is a $(\nu-1,\mu-1)$-solution of (E_1). Indeed, if $(Y(\lambda),Z(\lambda))$ is an arbitrary solution of (E_1) with $Z(\lambda) = \sum_{j=0}^{\gamma} \lambda^j z_j$, $\gamma \leq \mu$, set

$$\tilde{Z}(\lambda) = Z(\lambda) - \lambda^{\gamma-\mu} M(\lambda) m_\mu^{-1} z_\gamma, \qquad \tilde{Y}(\lambda) = Y(\lambda) + \lambda^{\gamma-\mu} m_\mu^{-1} z_\gamma L(\lambda).$$

Obviously, $(\tilde{Y}(\lambda),\tilde{Z}(\lambda))$ is a solution of (E_1) and $\deg\tilde{Z} \leq \gamma-1$. Proceeding in this way one reduces the degree of $Z(\lambda)$ to $\mu-1$. Now, if $(Y(\lambda),Z(\lambda))$ is a solution of (E_1) such that the estimation $\deg Z \leq \mu-1$ holds true, then necessarily $Y(\lambda) = \sum_{j=0}^{\delta} \lambda^j y_j$ with $\delta \leq \nu-1$. Indeed, assuming that $\delta \geq \nu$, $\det y_\delta \neq 0$, we have $\mu+\delta > \mu-1+\nu$, i.e., $\deg M(\lambda)Y(\lambda) > \deg Z(\lambda)L(\lambda)$. Hence $m_\mu y_\delta = 0$, which yields $y_\delta = 0$ in view of the invertibility of m_μ.

The main aim of this section is to prove the following criterion for solvability of (E_1).

THEOREM 4.1. *Let $M(\lambda)$ and $L(\lambda)$ be regular matrix polynomials and let m_μ be invertible. Then the equation (E_1) is solvable if and only if for every common eigenvalue $\lambda_0 \in \sigma(M) \cap \sigma(L)$ of M and L the following condition is fulfilled:*

$$\sum_{i=0}^{k} f_i R g_{k-i} = 0 \quad (k = 0,1,\ldots,\min(\alpha,\beta)) \tag{4.2}$$

for any left Jordan chain $f_0,\ldots,f_\alpha$ of $M(\lambda)$ and for any right Jordan chain $g_0,\ldots,g_\beta$ of $L(\lambda)$, corresponding to the eigenvalue λ_0.

Note that in (4.2) it suffices to take only the Jordan chains from the appropriate canonical systems.

We remark that the above result, in a more general setting of equations (E_1) with right hand side R depending on λ, is contained in an unpublished manuscript [16] of the present authors, where the proof is based on the authors' article [15] on tensor resultants (see also [12]). Here we present another proof of Theorem 4.1 which is divided into three steps. In the first two steps we consider the linear case, i.e., we assume that $M(\lambda) = \lambda I+A$ and $L(\lambda) = \lambda I+B$. In this case, in view of the remarks made at the beginning of this section, the solvability of (E_1) is equivalent to the solvability of the linear matrix equation of Lyaponov type:

$$AS - SB = R. \tag{E_2}$$

In the first step we assume that the matrices A and B in (E_2) are lower and upper one-cell Jordan matrices, respectively, with the same eigenvalue. Then in Step 2, the general linear matrix equation (E_2) is treated. Finally, in the third step we apply the connections between solvability of equations in matrix polynomials and solvability of Lyapunov type matrix equations, as developed in [24], [25], in order to reduce the general case of equation (E_1) to the case of a linear equation (E_2).

PROOF OF THEOREM 4.1.

Step 1. In this step we show that if

$$A = \underbrace{\begin{bmatrix} \lambda_0 & & & \\ 1 & \lambda_0 & & \\ & \ddots & \ddots & \\ & & 1 & \lambda_0 \end{bmatrix}}_{\alpha+1}, \qquad B = \underbrace{\begin{bmatrix} \lambda_0 & 1 & & \\ & \lambda_0 & \ddots & \\ & & \ddots & 1 \\ & & & \lambda_0 \end{bmatrix}}_{\beta+1} \tag{4.3}$$

are Jordan cells, then the equation

$$AS - SB = C := [c_{ij}]_{i,j=0}^{\alpha,\beta} \tag{E_3}$$

is solvable if and only if

$$\sum_{i=0}^{k} c_{i,k-i} = 0, \quad k = 0,1,\dots,\min\{\alpha.\beta\}. \tag{4.4}$$

Moreover, if (4.4) is fulfilled, then a solution $S = \tilde{S} = [\tilde{s}_{ij}]_{i,j=1}^{\alpha+1,\beta+1}$ of (E_3) is given as follows:

if $\alpha \geqslant \beta$, then for all $j = 1,2,\dots,\beta+1$ set

$$\tilde{s}_{ij} = \begin{cases} \sum\limits_{\ell=i}^{\min\{i+j-1,\alpha\}} c_{\ell,i+j-1-\ell}, & i = 1,\dots,\alpha \\ 0 & , \ i = \alpha+1 \end{cases} ; \tag{4.5}$$

if $\alpha \leqslant \beta$, then for all $i = 1,2,\dots,\alpha+1$ set

$$\tilde{s}_{ij} = \begin{cases} -\sum\limits_{\ell=j}^{\min\{i+j-1,\beta\}} c_{i+j-1-\ell,\ell} & , \ j = 1,\dots,\beta \\ 0 & , \ j = \beta+1 \end{cases} . \tag{4.6}$$

The general solution of (E_3) is of the form

$$S = \tilde{S} + H , \tag{4.7}$$

where $\tilde{S}$ is defined as above and $H = [h_{i+j-1}]_{i,j=1}^{\alpha+1,\beta+1}$ is an arbitrary Hankel matrix with

$$h_\ell = 0, \quad \ell = 1,\dots,\max\{\alpha,\beta\}. \tag{4.8}$$

To prove the above assertions we shall assume without loss of generality that $\lambda_0 = 0$ in (4.3). Then one easily sees that (E_3) is equivalent to the following system of equations:

$$c_{00} = 0, \tag{4.9}$$

$$s_{i1} = c_{i0} \ (i=1,\dots,\alpha), \tag{4.10}$$

$$s_{1j} = -c_{0j} \ (j=1,\dots,\beta), \tag{4.11}$$

$$s_{i,j+1} - s_{i+1,j} = c_{ij} \ (i=1,\dots,\alpha;\ j=1,\dots,\beta), \tag{4.12}$$

where $S = [s_{ij}]_{i,j=1}^{\alpha+1,\beta+1}$. Starting with the initial values (4.10) we obtain from the difference equations (4.12) the following formula for the entries s_{ij} with $i+j \leqslant \alpha+1$:

$$s_{ij} = \sum_{\ell=i}^{i+j-1} c_{\ell,i+j-1-\ell} \quad (i+j \leqslant \alpha+1). \tag{4.13}$$

On the other hand, starting with the initial values (4.11) and applying again the difference equations (4.12), we derive the following expressions for the entries s_{ij} with $i+j \leqslant \beta+1$:

$$s_{ij} = -\sum_{\ell=0}^{i-1} c_{\ell,i+j-1-\ell} \quad (i+j \leqslant \beta+1). \tag{4.14}$$

Comparing (4.13) and (4.14) we conclude that for all i,j such that $i+j \leqslant \min(\alpha,\beta)+1$, the condition

$$\sum_{\ell=0}^{i+j-1} c_{\ell,i+j-1-\ell} = 0 \quad \left(i+j \leqslant \min(\alpha,\beta)+1\right) \tag{4.15}$$

is necessary for solvability of (4.9)-(4.12), and hence for solvability of (E_3). It remains to note that (4.15) is precisely the same as (4.4).

Now assume that condition (4.4) is fulfilled and let us show that the matrix $\tilde{S}$ defined by (4.5)-(4.6) is a solution of (E_3). Consider, for example, the case $\alpha \geqslant \beta$. Setting $j = 1$ in (4.5) we see that

$$\tilde{s}_{i1} = c_{i0} \quad (i=1,\dots,\alpha). \tag{4.16}$$

Furthermore, formulas (4.5) imply that

$$\tilde{s}_{i,j+1} - \tilde{s}_{i+1,j} = \sum_{\ell=i}^{\min(i+j,\alpha)} c_{\ell,i+j-\ell} - \sum_{\ell=i+1}^{\min(i+j,\alpha)} c_{\ell,i+j-\ell} = c_{ij}, \tag{4.17}$$

for any $j = 1,\dots,\beta$ and for $i = 1,\dots,\alpha-1$, while

$$\tilde{s}_{\alpha,j+1} - \tilde{s}_{\alpha+1,j} = \tilde{s}_{\alpha,j+1} = c_{\alpha,j} \quad (j=1,\dots,\beta). \tag{4.18}$$

Using the condition (4.4) we also obtain from (4.5) that

$$\tilde{s}_{1j} = \sum_{\ell=1}^{j} c_{\ell,j-\ell} = -c_{oj}. \tag{4.19}$$

Equalities (4.16)-(4.19) show that the entries of the matrix $\tilde{S}$ satisfy the equations (4.10)-(4.12) which are equivalent to the equations (E_3). (Note that (4.9) is contained among the equalities (4.4)). So, the matrix $\tilde{S}$ is a solution of (E_3). The case $\beta \geqslant \alpha$ is treated in a similar way (or by passing in (E_3) to the transposed matrices).

To prove formula (4.7) we note that if $C = 0$ in (E_3), then any solution of the homogeneous) equation (E_3) is a Hankel matrix $S = H = [h_{i+j-1}]_{i,j=1}^{\alpha+1,\beta+1}$ (see e.g., [21]), and formulas (4.13)-(4.14) imply that in this case (4.8) holds true.

Step 2. In this step we prove that equation (E_2) with $A \in \mathbb{C}^{p\times p}$, $B \in \mathbb{C}^{q\times q}$ and $R \in \mathbb{C}^{p\times q}$ is solvable if and only if for any common eigenvalue $\lambda_0 \in \sigma(A) \cap \sigma(B)$ the following condition is fulfilled:

$$\sum_{i=0}^{k} x_i R y_{k-i} = 0 \quad (k=0,1,\ldots,\min\{\alpha,\beta\}) \tag{4.20}$$

for any left Jordan chain $x_0,\ldots,x_\alpha$ of A and for any right Jordan chain $y_0,\ldots,y_\beta$ of B, corresponding to λ_0.

To this end we represent

$$A = X^{-1} J_A^{(\ell)} X, \qquad B = Y J_B^{(u)} Y^{-1},$$

where $J_A^{(\ell)}$ and $J_B^{(u)}$ are lower and upper Jordan matrices, respectively, and rewrite (E_2) in the form

$$J_A^{(\ell)}\tilde{S} - \tilde{S}J_B^{(u)} = XRY, \tag{E_4}$$

where $\tilde{S} = XSY$. Furthermore, represent

$$\begin{aligned} J_A^{(\ell)} &= J^{(\ell)}(\lambda_1) \dotplus J^{(\ell)}(\lambda_2) \dotplus \ldots \dotplus J^{(\ell)}(\lambda_t), \\ J_B^{(r)} &= J^{(r)}(\mu_1) \dotplus J^{(r)}(\mu_2) \dotplus \ldots \dotplus J^{(r)}(\mu_s); \end{aligned} \tag{4.21}$$

where $J^{(\ell)}(\lambda_j)(J^{(r)}(\mu_j))$ denotes an elementary lower (upper) Jordan cell corresponding to the eigenvalue λ_j (μ_j). Make the corresponding partitionings in the matrices X and Y:

$$X = \begin{bmatrix} X_1 \\ X_2 \\ \vdots \\ X_t \end{bmatrix}, \quad Y = [Y_1 Y_2 \dots Y_s], \tag{4.22}$$

and represent $\tilde{S}$ as a block matrix $\tilde{S} = [\tilde{S}_{ij}]_{i,j=1}^{t,s}$, where the number of rows (columns) in the block $\tilde{S}_{ij}$ is equal to the size of the matrix $J^{(\ell)}(\lambda_i)(J^{(r)}(\lambda_j))$. With these notations, equation (E_4) can be rewritten as the following system of equations

$$J^{(\ell)}(\lambda_k)\tilde{S}_{ki} - \tilde{S}_{ki}J^{(r)}(\mu_i) = X_k R Y_i \quad (k=1,\dots,t;\ \ i=1,\dots,s). \tag{4.23}$$

As is well known, all equations (4.23) with $\lambda_k \neq \mu_i$ are solvable. Now assume that $\lambda_0 := \lambda_{k_0} = \mu_{i_0}$ is a common eigenvalue of A and B. Represent

$$X_{k_0} = \begin{bmatrix} x_0 \\ x_1 \\ \vdots \\ x_\alpha \end{bmatrix}, \quad Y_{i_0} = [y_0 y_1 \dots y_\beta],$$

and rewrite (4.23) for $i = i_0$ and $k = k_0$ in the form

$$J^{(\ell)}(\lambda_0)\tilde{S}_{k_0 i_0} - \tilde{S}_{k_0 i_0}J^{(r)}(\lambda_0) = [x_p R y_q]_{p,q=0}^{\alpha,\beta}. \tag{4.24}$$

It follows from Step 1 that (4.24) is solvable if and only if condition (4.20) is fulfilled.

Step 3. In this step we prove that condition (4.2) is necessary and sufficient for solvability of (E_1). To this end we shall use the results of [24],[25] which establish the connections between the solutions of equation (E_1) and an associated matrix equation of type (E_2). We present now a specification of these results which is needed for our purposes.

THEOREM 4.2. *Let* (V,Ψ) *and* (Φ,U) *be finite left and right Jordan pairs of the regular matrix polynomials* $M(\lambda)$ *and* $L(\lambda)$, *respectively. Assume that* m_μ, *the leading coefficient of* $M(\lambda)$, *is invertible. Then equation* (E_1) *is solvable if and only if the equation*

$$VS - SU = -\Psi R\Phi \qquad (E_5)$$

is solvable.

Now let (V,Ψ) and (Φ,U) be as in Theorem 4.2 and apply the result of Step 2 to the equation (E_5). We infer that equation (E_1) is solvable if and only if

$$\sum_{j=0}^{k} x_j \Psi R\Phi y_{k-j} = 0 \quad \left(k=0,1,\ldots,\min\{\alpha,\beta\}\right) \qquad (4.25)$$

for any left Jordan chain $x_0,\ldots,x_\alpha$ of V and for any right Jordan chain $y_0,\ldots,y_\beta$ of U, corresponding to a common eigenvalue $\lambda_0 \in \sigma(U) \cap \sigma(V)$. It remains to note that if $y_0,\ldots,y_\beta$ is a right Jordan chain of U corresponding to $\lambda_0 \in \sigma(U)$, then $\Phi y_0,\ldots,\Phi y_\beta$ is a right Jordan chain of $L(\lambda)$ corresponding to the eigenvalue $\lambda_0 \in \sigma(L)$. Moreover, by taking all Jordan chains $y_0,\ldots,y_\beta$ from the Jordan basis of U, one obtains the canonical system of right Jordan chains of the polynomial $L(\lambda)$. Similarly, the canonical system of left Jordan chains of $M(\lambda)$ consists of sequences $x_0\Psi,\ldots,x_\alpha\Psi$, where the sequence $\{x_i\}_{i=0}^{\alpha}$ runs over all left Jordan chains from the left Jordan basis of V. Thus, equalities (4.25) can be written as (4.2), which completes the proof of the Theorem. □

Note that after the first version of this paper was completed, a paper of J. Ball and A. Ran [3] appeared, where the result of Step 2 is established for the case R = I (see Lemma 4.2). Note also that formulas (4.5)-(4.6), along with well known formulas for the case $\sigma(A) \cap \sigma(B) = \phi$ (see, e.g., [6]), imply explicit formulas for the solutions of (E_2) in terms of the Jordan chains of A and B. These formulas are not needed for the purposes of the present paper, and therefore are omitted.

In Section 7 we shall need also a criterion for solvability of matrix equations of Stein type. First we prove the following result.

PROPOSITION 4.3. *Let $M(\lambda)$ and $L(\lambda)$ be regular matrix polynomials given by (4.1) such that the coefficients m_μ and ℓ_0 are invertible, and set $\hat{M}(\lambda):=m_\mu^{-1}M(\lambda)$, $\tilde{L}^0(\lambda):=\sum_{j=0}^{\nu}\lambda^j\ell_{\nu-j}\ell_0^{-1}$. Then the equation (E_1) is solvable if and only if the matrix equation*

$$X-C_{\hat{M}}XC_{\tilde{L}^0}=\operatorname{diag}\left(m_\mu^{-1}R\ell_0^{-1},0,\dots,0\right) \tag{E_6}$$

is solvable.

PROOF. First recall that the finite right Jordan pair of the polynomial $L(\lambda)\ell_0^{-1}$ can be obtained from the companion right standard pair $(\Phi^{(0)},C_{\tilde{L}^0})$ of the monic polynomial $\tilde{L}^0(\lambda)$ in the following manner: Decompose the space $\mathbb{C}^{\nu r}$ into a direct sum $\mathbb{C}^{\nu r}=M\dot{+}N$ so that $\sigma(C_{\tilde{L}^0}|N)=\{0\}$ and $0\notin\sigma(C_{\tilde{L}^0}|M)$, and represent the pair $(\Phi^{(0)},C_{\tilde{L}^0})$ with respect to this decomposition:

$$C_{\tilde{L}^0}=\begin{bmatrix}C_1 & 0\\ 0 & C_0\end{bmatrix},\quad \Phi^{(0)}=[\Phi_1^{(0)}\Phi_0^{(0)}]\,. \tag{4.26}$$

Then $(\Phi_1^{(0)},C_1^{-1})$ is a finite right Jordan pair of $L(\lambda)\ell_0^{-1}$ as well as the pair $(\Phi_1^{(0)}C_1^{-1},C_1^{-1})$, which is similar to the former.

By applying Theorem 4.2 to the equation

$$\hat{M}(\lambda)Y(\lambda)+Z(\lambda)L(\lambda)\ell_0^{-1}=m_\mu^{-1}R\ell_0^{-1}, \tag{$\tilde{E}_1$}$$

which is equivalent to (E_1), we see that the equation (E_1) is solvable if and only if the matrix equation

$$C_{\hat{M}}S-SC_1^{-1}=-\Psi^{(0)}m_\mu^{-1}R\ell_0^{-1}\Phi_1{}^{(0)}C_1{}^{-1}$$

is solvable.

Now rewrite the last equation in the form

$$S-C_{\hat{M}}SC_1=\Psi^{(0)}m_\mu^{-1}R\ell_0^{-1}\Phi_1^{(0)}, \tag{4.27}$$

and consider the equation

$$S_0 - \hat{C}_{\hat{M}} S_0 C_0 = \Psi^{(0)} m_\mu^{-1} R \ell_0^{-1} \Phi_0^{(0)} , \tag{4.28}$$

where C_0 and $\Phi_0^{(0)}$ come from the representation (4.26). The equation (4.28) is always solvable since $\lambda\mu \neq 1$ for any $\lambda \in \sigma(\hat{C}_{\hat{M}})$ and $\mu \in \sigma(C_0) = \{0\}$. Using the representation (4.26) and representing $X = [S\ S_0]$, one sees that the equation (E_6) can be written as the pair of equations (4.27)-(4.28), and hence it is solvable if and only if the equation (4.27) is solvable. □

We now transform the equation (E_6) to a form which is convenient for our purposes. Firstly, we use (2.4) to rewrite (E_6) in the form

$$Y - C_{\hat{M}} Y \hat{C}_{\tilde{L}O} = S_{\hat{M}}^{-1} \Psi^{(0)} m_\mu^{-1} R \ell_0^{-1} \Phi^{(0)} S_{\tilde{L}O} , \tag{4.29}$$

with $Y = S_{\hat{M}}^{-1} X S_{\tilde{L}O}^{-1}$. Secondly, substitute (2.3) to obtain

$$S - \hat{K}_{\hat{M}} S K_{\tilde{L}O} = (S_{\hat{M}} P)^{-1} \Psi^{(0)} m_\mu^{-1} R \ell_0^{-1} \Phi^{(0)} (P S_{\tilde{L}O})^{-1} ,$$

with $S = PYP$. A simple calculation shows that the right hand side in the last equation is equal to $\Psi^{(0)} m_\mu^{-1} R \ell_0^{-1} \Phi^{(0)}$, and hence the equation (E_6) is transformed to the equation

$$S - \hat{K}\hat{M} S K \tilde{L}^{O} = \operatorname{diag}(m_\mu^{-1} R \ell_0^{-1}, 0, \ldots, 0) , \tag{E_7}$$

with $S = P S_{\hat{M}}^{-1} X S_{\tilde{L}}^{-1} P$.

Now Theorem 4.1 and Proposition 4.3 imply the following result.

THEOREM 4.4. *Let $M(\lambda)$ and $L(\lambda)$ be as in Proposition 4.3. Then the following are equivalent:*

(a) the equation (E_1) is solvable;

(b) the equation (E_7) is solvable;

(c) for every $\lambda_0 \in \sigma(M) \cap \sigma(L)$ and for any left Jordan chain $f_0, \ldots, f_\alpha$ of $M(\lambda)$ and any right Jordan chain $g_0, \ldots, g_\beta$ of $L(\lambda)$ corresponding to λ_0, the equation

$$\sum_{i=0}^{k} f_i R g_{k-i} = 0 \quad (k=0,1,\ldots,\min(\alpha,\beta))$$

holds true.

5. COPRIME SYMMETRIC FACTORIZATIONS

Let $F(\lambda) = \sum_{j=-n}^{n} \lambda^j f_j$ be a rxr rational matrix polynomial that is nonnegative on the unit circle $\Gamma_0 = \{\lambda \in \mathbb{C} \mid |\lambda|=1\}$, i.e., $\langle F(\lambda)g,g\rangle \geq 0$ for any $\lambda \in \Gamma_0$ and for any $g \in \mathbb{C}^r$. A *left symmetric factorization of degree* n of $F(\lambda)$ is defined as the representation

$$F(\lambda) = [A(\lambda)]^* A(\lambda) \quad (\lambda \in \Gamma_0), \tag{5.1}$$

where $A(\lambda) = \sum_{j=0}^{n} \lambda^j a_j$ is a matrix polynomial of degree n whose leading coefficient a_n is positive definite: $a_n > 0$. Similarly, a *right symmetric factorization of degree* n of $F(\lambda)$ is defined as the representation

$$F(\lambda) = B(\lambda)[B(\lambda)]^* \quad (\lambda \in \Gamma_0), \tag{5.2}$$

where $B(\lambda) = \sum_{j=0}^{n} \lambda^j b_j$ is a matrix polynomial of degree n with positive leading coefficient $b_n > 0$. It is known (see e.g., [14]) that a nonnegative rational polynomial $F(\lambda)$ always admits right and left symmetric factorizations.

Extending the factorization (5.1)-(5.2) from the unit circle to the complex plane, we can write

$$F(\lambda) = \lambda^{-n} A^{\circledast}(\lambda) A(\lambda) \quad (\lambda \in \mathbb{C}) \tag{5.1'}$$

and

$$F(\lambda) = \lambda^{-n} B(\lambda) B^{\circledast}(\lambda) \quad (\lambda \in \mathbb{C}), \tag{5.2'}$$

respectively, where $A^{\circledast}(\lambda) := \sum_{j=0}^{n} \lambda^j a_{n-j}^*$. Given left and right symmetric factorizations (5.1) and (5.2) of $F(\lambda)$, we say that the factorizations are *relatively coprime* if the polynomials $A(\lambda)$ and $B^{\circledast}(\lambda)$ are right coprime or, which is the same, $A^{\circledast}(\lambda)$ and $B(\lambda)$ are left coprime.

In this section we deal with the following problem: given a left symmetric factorization (5.1) of order n of $F(\lambda)$, find a right symmetric factorization (5.2) of order n of $F(\lambda)$ that is relatively coprime with (5.1). A solution of this problem is known for the case where the eigenvalues of $A(\lambda)$ lie inside the unit circle. Indeed, rewriting (5.1) in the form

$$F(\lambda) = [L(\lambda)]^* L(\lambda) \qquad (\lambda \in \Gamma_0), \tag{5.3}$$

with $L(\lambda) := \lambda^{-n} A(\lambda)$, we see that in this case (5.3) is a left canonical Wiener-Hopf factorization of $F(\lambda)$ with respect to the unit circle. Then well known results (see e.g., [5],[14]) ensure the existence of a right canonical factorization of $F(\lambda)$:

$$F(\lambda) = R(\lambda)[R(\lambda)]^* \qquad (\lambda \in \Gamma_0), \tag{5.3'}$$

where $R(\lambda)$ is of the form $R(\lambda) = \sum_{j=0}^{n} \lambda^{-j} r_j$, $r_0 > 0$, and $R(\lambda)$ is invertible for any λ outside the unit circle. It is clear that setting $B(\lambda) = \lambda^n R(\lambda)$ we can rewrite (5.3') in the form (5.2) and the polynomial $B^{\circledast}(\lambda)$, whose eigenvalues lie outside the unit circle, is coprime with $A(\lambda)$. Moreover, as follows from our results (see Corollary 8.3) the above choice of $B(\lambda)$ is unique in the case under consideration. This is not true in general as we shall see in the present section and in Section 8.

Before turning to the general setting of the problem, consider another particular case. Namely, assume that $A(\lambda) = \sum_{j=0}^{n} a_j \lambda^j$ in (5.1) is just a scalar polynomial. Obviously, in this case we can represent $F(\lambda)$ in the form (5.2) with $B(\lambda) = A(\lambda)$. It is clear then that $A(\lambda)$ and $B^{\circledast}(\lambda) = \sum_{j=0}^{n} \bar{a}_j \lambda^{n-j}$ are coprime if and only if $A(\lambda)$ does not vanish on the unit circle and does not have pairs of roots that are symmetric with respect to the unit circle. A simple analysis shows that the above condition on the zeroes of $A(\lambda)$ is also necessary for existence of a right symmetric factorization (5.2) of order n that is relatively coprime with (5.1). The theorem which follows shows that in the case of polynomials with matrix coefficients, pairs of symmetric eigenvalues of $A(\lambda)$ may occur, but then some geometric relations between the corresponding Jordan chains have to be fulfilled.

THEOREM 5.1. *Let* $A(\lambda) = \sum_{j=0}^{n} \lambda^j a_j$ *be an rxr matrix polynomial with positive definite leading coefficient:* $a_n > 0$. *Then the function*

$$F(\lambda) := [A(\lambda)]^* A(\lambda) \qquad (\lambda \in \Gamma_0) \tag{5.4}$$

admits a right symmetric factorization of order n *that is relatively coprime with* (5.4) *if and only if for every symmetric pair of eigenvalues* $\lambda_0, \bar{\lambda}_0^{-1}$ *of* $A(\lambda)$ *(if any) the following conditions are fulfilled:*

$$\sum_{i=0}^{k} \langle \phi_i, \psi_{k-i} \rangle = 0 \qquad (k=0,1,\dots,\min(\alpha,\beta)) \tag{5.5}$$

for any left Jordan chains $\phi_0, \phi_1, \dots, \phi_\alpha$ *and* $\psi_0, \psi_1, \dots, \psi_\beta$ *of* $A(\lambda)$ *corresponding to* λ_0 *and* $\bar{\lambda}_0^{-1}$, *respectively.*

PROOF. We first prove the necessity. Assume that there exists a matrix polynomial $B(\lambda)$ of degree n with positive leading coefficient such that

$$F(\lambda) = B(\lambda)[B(\lambda)]^* \qquad (\lambda \in \Gamma_0), \tag{5.6}$$

and the polynomials $A(\lambda)$ and $B^{\circledast}(\lambda)$ are right coprime. It follows that there are matrix polynomials $\Gamma(\lambda)$ and $\Delta(\lambda)$ such that

$$\Delta(\lambda) B^{\circledast}(\lambda) + \Gamma(\lambda) A(\lambda) = I. \tag{5.7}$$

Since $A^{\circledast}(\lambda)$ and $B(\lambda)$ are left coprime, there are matrix polynomials $\Omega(\lambda)$ and $\Xi(\lambda)$ such that

$$B(\lambda)\Omega(\lambda) + A^{\circledast}(\lambda)\Xi(\lambda) = I. \tag{5.8}$$

Rewriting (5.6) in the form

$$\lambda^{-n} A^{\circledast}(\lambda) A(\lambda) = \lambda^{-n} B(\lambda) B^{\circledast}(\lambda),$$

and combining this with (5.7)-(5.8), we have

$$\begin{bmatrix} \Delta(\lambda) & \Gamma(\lambda) \\ B(\lambda) & -A^{\circledast}(\lambda) \end{bmatrix} \begin{bmatrix} B^{\circledast}(\lambda) & \Omega(\lambda) \\ A(\lambda) & -\Xi(\lambda) \end{bmatrix} = \begin{bmatrix} I & C(\lambda) \\ 0 & I \end{bmatrix},$$

where $C(\lambda) := \Delta(\lambda)\Omega(\lambda) - \Gamma(\lambda)\Xi(\lambda)$. Multiplying this equality by $\begin{bmatrix} I & -C(\lambda) \\ 0 & I \end{bmatrix}$ from the right we obtain

$$\begin{bmatrix} \Delta(\lambda) & \Gamma(\lambda) \\ B(\lambda) & -A^{\circledast}(\lambda) \end{bmatrix} \begin{bmatrix} B^{\circledast}(\lambda) & \tilde{\Omega}(\lambda) \\ A(\lambda) & -\tilde{\Xi}(\lambda) \end{bmatrix} = \begin{bmatrix} I & 0 \\ 0 & I \end{bmatrix}, \tag{5.9}$$

where $\tilde{\Omega}(\lambda)$ and $\tilde{\Xi}(\lambda)$ are some matrix polynomials. Interchanging the order of the factors in (5.9) we see, in particular, that

$$A(\lambda)\Gamma(\lambda) + \tilde{\Xi}(\lambda)A^{\circledast}(\lambda) = I. \tag{5.10}$$

Now Theorem 4.1 implies that if $\lambda_0 \in \sigma(A) \cap \sigma(A^{\circledast})$, then the conditions

$$\sum_{i=0}^{k} \phi_i g_{k-i} = 0 \quad (k=0,1,\dots,\min(\alpha,\beta)) \tag{5.11}$$

are fulfilled for any right Jordan chain $g_0,\dots,g_\beta$ of $A^{\circledast}(\lambda)$ and any left Jordan chain $\phi_0,\dots,\phi_\alpha$ of $A(\lambda)$ corresponding to λ_0. Note that $\lambda_0 \in \sigma(A) \cap \sigma(A^{\circledast})$ if and only if $\lambda_0, \bar{\lambda}_0^{-1} \in \sigma(A)$ and that $g_0,\dots,g_\beta$ is a right Jordan chain of $A^{\circledast}(\lambda)$ corresponding to λ_0 if and only if $\bar{g}_0, \bar{g}_1,\dots,\bar{g}_\beta$ is a left Jordan chain of $A(\lambda)$ corresponding to $\bar{\lambda}_0^{-1}$. (Here and elsewhere $\bar{g}$ denotes the vector whose j^{th} coordinate is the complex conjugate of the jth cordinate of g). Thus the equalities (5.11) become (5.5) and the necessity is proved.

Passing to the proof of sufficiency, we first observe that in view of Theorem 4.1, the preceding paragraph shows also that conditions (5.5) imply the existence of matrix poynomials $\Gamma(\lambda)$ and $\tilde{\Xi}(\lambda)$ which solve the equation (5.10). However, for our purposes this solution has to be symmetrized in a certain sense.

We need the following notations. If $M(\lambda) = \sum_{j=0}^{n} \lambda^j m_j$ is a matrix polynomial, we set

$$M^{[]}(z) := (1-iz)^n M\left(\frac{1+iz}{1-iz}\right) = \sum_{j=0}^{n} (1-iz)^{n-j}(1+iz)^j m_j,$$

$$M_{[]}(z) := (1+z)^n M\left(\frac{1-z}{1+z}i\right) = \sum_{k=0}^{n} (1+z)^{n-k}[(1-z)i]^k m_k.$$

It is clear that $(M^{[]})_{[]} = 2^n M$ and $(M_{[]})^{[]} = 2^n M$. Furthermore, we denote for brevity, $M^{[*]} = (M^{[]})^*$, i.e.,

$$M^{[*]}(z) := \sum_{j=0}^{n} (1+iz)^{n-j}(1-iz)^j m_j^*,$$

and we note that

$$(M^{\circledast})^{[]}(z) := (1-iz)^n M^{\circledast}\left(\frac{1+iz}{1-iz}\right) = \sum_{j=0}^{n} (1-iz)^{n-j}(1+iz)^j m_{n-j}^* = M^{[*]}(z).$$

Proceeding with the proof of sufficiency, we remark that $z_0 = -i \notin \sigma(A^{[]})$ and $z_0 = i \notin \sigma(A^{[*]})$, since $A^{[]}(-i) = 2^n a_n$, $A^{[*]}(i) = 2^n a_n^*$, and the matrix a_n is assumed to be positive definite. So, if $z_0 \in \sigma(A^{[]}) \cap \sigma(A^{[*]})$, then $z_0 \neq \pm i$, and hence $\lambda_0, \bar{\lambda}_0^{-1} \in \sigma(A)$, where $\lambda_0 = \frac{1+iz_0}{1-iz_0}$. In addition, if the sequence $\phi_0, \ldots, \phi_\alpha$ (resp., $\psi_0, \ldots, \psi_\beta$) is a left (resp., right) Jordan chain of $A^{[]}$ (resp., $A^{[*]}$), corresponding to z_0, then one easily checks that $\phi_0, \ldots, \phi_\alpha$ (resp., $\bar{\psi}_0, \ldots, \bar{\psi}_\beta$) is a left Jordan chain of A corresponding to λ_0 (resp., $\bar{\lambda}_0^{-1}$). Thus, in view of Theorem 4.1, condition (5.5) yields the existence of matrix polynomials $\Phi(z)$ and $\Psi(z)$ such that

$$A^{[]}(z)\Phi(z) + \Psi(z)A^{[*]}(z) = I. \tag{5.12}$$

Now observe that condition (5.5) implies that $A(\lambda)$ has no eigenvalues on the unit circle. Since the leading coefficient of $A^{[]}(z)$ is equal to $(-i)^n A(-1)$, it follows, in particular, that $A^{[]}(z)$ and $A^{[*]}(z)$

have invertible leading coefficients. Hence, in view of the observations made at the beginning of Section 4, we may assume that $\deg\Phi \leqslant n-1$, $\deg\Psi \leqslant n-1$ for the solution of (5.12). Furthermore, one can find a solution of (5.12) which, in addition, satisfies the condition $\Phi(\lambda) = \Psi^*(\lambda)$. Indeed, if $\Phi_1(\lambda)$, $\Psi_1(\lambda)$ is a solution of (5.12), then obviously, the pair of polynomials $\Phi = \frac{1}{2}(\Phi_1 + \Psi_1^*)$, $\Psi = \frac{1}{2}(\Phi_1^* + \Psi_1)$ forms a solution of (5.12) as well.

So let $\Phi(\lambda)$ be a polynomial of degree $\leqslant n-1$ such that

$$A^{[\,]}(z)\Phi(z) + \Phi^*(z)A^{[*]}(z) = I, \tag{5.13}$$

and introduce the rational matrix function $W(\lambda) := [A^{[\,]}(\lambda)]^{-1}\Phi^*(\lambda)$. Equation (5.13) can be rewritten as

$$W(\lambda) + W^*(\lambda) = [A^{[\,]}(\lambda)]^{-1}[A^{[*]}(\lambda)]^{-1}, \tag{5.14}$$

where, by definition, $W^*(\lambda) := [W(\bar{\lambda})]^*$. Substituting a right coprime matrix fraction representation of W — say, $W = \Theta(\lambda)Z^{-1}(\lambda)$ — in (5.14), we obtain

$$Z^*(\lambda)\Theta(\lambda) + \Theta^*(\lambda)Z(\lambda) = Z^*(\lambda)[A^{[\,]}(\lambda)]^{-1}[A^{[*]}(\lambda)]^{-1}Z(\lambda) \overset{\text{def}}{=} V(\lambda). \tag{5.15}$$

From (5.13) we know that the matrix fraction $W(\lambda) = [A^{[\,]}(\lambda)]^{-1}\Phi^*(\lambda)$ is coprime. Since the matrix fraction $W(\lambda) = \Theta(\lambda)Z^{-1}(\lambda)$ is also coprime, we infer that $\det Z(\lambda) = \det A^{[\,]}(\lambda)$ and $\det Z^*(\lambda) = \det A^{[*]}(\lambda)$. It follows that $\det V(\lambda) \equiv 1$, and since the left hand side of (5.15) is a polynomial, we conclude that $V(\lambda)$ is a unimodular polynomial.

We know already that $A(\lambda)$ does not have eigenvalues on the unit circle. This implies that $A^{[\,]}(\lambda)$ (and hence $Z(\lambda)$) does not have eigenvalues on the real line. So, for any real λ and any $\phi \in \mathbb{C}^r$ ($\phi \neq 0$) we have

$$\langle V(\lambda)\phi,\phi\rangle = \|[A^{[*]}(\lambda)]^{-1}Z(\lambda)\phi\|^2 > 0,$$

i.e., the unimodular matrix polynomial $V(\lambda)$ is positive on the real line. We claim that there is a unimodular matrix polynomial $U(\lambda)$ such that $V(\lambda) = U^*(\lambda)U(\lambda)$. Indeed, let k be an integer such that $2k \geqslant \deg V(\lambda)$, and introduce $L(\lambda) = \lambda^{2k}[V(0)]^{-\frac{1}{2}}V(\frac{1}{\lambda})[V(0)]^{-\frac{1}{2}}$. Then $L(\lambda)$ is a monic polynomial of degree 2k with hermitian coefficients and such that $\langle L(\lambda)\phi,\phi\rangle \geqslant 0$ for any real λ and any $\phi \in \mathbb{C}^r$. Hence, in view of Theorem II.2.6 in [14], there is a monic polynomial $M(\lambda)$ of degree k such that $L(\lambda) = M^*(\lambda)M(\lambda)$. Setting $U(\lambda) := \lambda^k M(\lambda^{-1})[V(0)]^{\frac{1}{2}}$, we obtain the factorization

$$V(\lambda) = U^*(\lambda)U(\lambda). \tag{5.16}$$

Note that $\lambda = 0$ is the only eigenvalue of $L(\lambda)$, and hence of $M(\lambda)$. It follows that the only possible eigenvalue of $U(\lambda)$ is $\lambda = 0$. But $U(0) = [V(0)]^{\frac{1}{2}}$, which is an invertible matrix, and therefore $\lambda = 0$ is not an eigenvalue of $U(\lambda)$. We see that $\det U(\lambda) = \text{const}$, and since $|\det U(\lambda)| = \det V(\lambda) = 1$, we infer that $\det U(\lambda) \equiv 1$. So, (5.16) provides the desired factorization of $V(\lambda)$.

Now set

$$\tilde{Z}(\lambda) := Z(\lambda)[U(\lambda)]^{-1}, \quad \tilde{\Theta}(\lambda) = \Theta(\lambda)[U(\lambda)]^{-1},$$

and use (5.16) to rewrite (5.15) in the form

$$\tilde{Z}^*(\lambda)\tilde{\Theta}(\lambda) + \tilde{\Theta}^*(\lambda)\tilde{Z}(\lambda) = I. \tag{5.17}$$

Clearly,

$$W(\lambda) := [A^{[]}(\lambda)]^{-1}\Phi^*(\lambda) = \tilde{\Theta}(\lambda)\tilde{Z}^{-1}(\lambda). \tag{5.18}$$

Combining (5.13), (5.17) and (5.18), we obtain

$$\begin{bmatrix} A^{[]}(\lambda) & \Phi^*(\lambda) \\ \tilde{Z}^*(\lambda) & -\tilde{\Theta}^*(\lambda) \end{bmatrix} \begin{bmatrix} \Phi(\lambda) & \tilde{\Theta}(\lambda) \\ A^{[*]}(\lambda) & -\tilde{Z}(\lambda) \end{bmatrix} = \begin{bmatrix} I & 0 \\ 0 & I \end{bmatrix}.$$

Interchanging the order of the factors, we have

$$\begin{bmatrix} \Phi(\lambda) & \tilde{\Theta}(\lambda) \\ A^{[*]}(\lambda) & -\tilde{Z}(\lambda) \end{bmatrix} \begin{bmatrix} A^{[]}(\lambda) & \Phi^{*}(\lambda) \\ \tilde{Z}^{*}(\lambda) & -\tilde{\Theta}^{*}(\lambda) \end{bmatrix} = \begin{bmatrix} I & 0 \\ 0 & I \end{bmatrix},$$

which implies, in particular, the relations

$$\tilde{Z}(\lambda)\tilde{Z}^{*}(\lambda) = A^{[*]}(\lambda)A^{[]}(\lambda) , \tag{5.19}$$

$$A^{[*]}(\lambda)\Phi^{*}(\lambda) + \tilde{Z}(\lambda)\tilde{\Theta}^{*}(\lambda) = I. \tag{5.20}$$

Observe that $\tilde{Z}(\lambda)$ is a polynomial of degree n with invertible leading coefficient. Indeed, if $\tilde{Z}(\lambda) = \sum_{j=0}^{m} \lambda^j z_j$ and $m > n$, then it is clear from (5.19) that $z_m z_m^* = 0$, i.e., $z_m = 0$, and hence we can set $m = n$.

Furthermore, $z_n z_n^* = \alpha_n^* \alpha_n$ where α_n is the leading coefficient of $A^{[]}(\lambda)$. We know already that α_n is invertible, and hence z_n is invertible as well.

Now introduce the polynomial

$$R(\lambda) = 2^n \tilde{Z}_{[]}(\lambda) = 2^n (1+\lambda)^n \tilde{Z}(\tfrac{1-\lambda}{1+\lambda} i),$$

and note that $(2^n \tilde{Z}^*)_{[]} = R^{\circledast}$. Substituting in (5.19) $\lambda = \frac{1-z}{1+z} i$ and multiplying both sides by $2^{2n}(1+z)^{2n}$ we obtain the equality

$$R(z)R^{\circledast}(z) = A^{\circledast}(z)A(z). \tag{5.21}$$

Furthermore, equality (5.20) shows that

$$\operatorname{Ker}A(z) \cap \operatorname{Ker}R^{\circledast}(z) = (0) \tag{5.22}$$

for any z except, perhaps, for $z = -1$. But in view of (5.5) $\operatorname{Ker}A(-1) = (0)$, and hence (5.22) holds true for any $z \in \mathbb{C}$, i.e., the polynomials $A(z)$ and $R^{\circledast}(z)$ are right coprime.

It remains to show that the polynomial $R(z)$ in (5.21) can be normalized so that its leading coefficient becomes positive definite. To this end observe that the leading coefficient r_n of $R(z)$ is invertible. Indeed, $r_n = 2^n\tilde{Z}(-i)$. But the matrix $A^{[\,]}(-i) = 2^n a_n$ is invertible and we know that the spectrum of $\tilde{Z}(\lambda)$ coincides with the spectrum of $A^{[\,]}(\lambda)$, which yields the invertibility of the matrix $\tilde{Z}(-i)$. Now set $B(z) := R(z)r_n^*(r_n r_n^*)^{-1/2}$. The leading coefficient of $B(z)$ is $(r_n r_n^*)^{1/2} > 0$, and (5.21) can be rewritten in the form

$$B(z)B^{\circledast}(z) = A^{\circledast}(z)A(z),$$

which completes the proof.

□

Analysing the above proof, one sees that the actual construction of the factor $B(\lambda)$ in a right symmetric factorization (5.2), that is relatively coprime with the left factorization (5.1), consists of two main steps, provided $A(\lambda)$ satisfies (5.5).

Step 1. Find a polynomial $\Phi(\lambda)$, with $\deg\Phi \leqslant n-1$, such that

$$A^{[\,]}(\lambda)\Phi(\lambda) + \Phi^*(\lambda)A^{[*]}(\lambda) = I. \tag{5.23}$$

Step 2. Find a polynomial $Z(\lambda)$ of degree n, with invertible leading coefficient, and a polynomial $\Theta(\lambda)$ of degree $\leqslant n-1$ such that

$$A^{[\,]}(\lambda)\Theta(\lambda) = \Phi^*(\lambda)Z(\lambda), \tag{5.24}$$

$$Z^*(\lambda)\Theta(\lambda) + \Theta^*(\lambda)Z(\lambda) = I, \tag{5.25}$$

and set

$$B(\lambda) := 2^n Z_{[\,]}(\lambda)Z(-i)\left[Z(-i)Z^*(i)\right]^{-1/2}.$$

We shall show now that the main ingredient in each of the above steps is solving a matrix-vector equation whose coefficient matrix is nicely structured.

Indeed, as follows from the results of [25], if $S = [s_{jk}]_{j,k=0}^{n-1}$ is a solution of the matrix equation

$$\hat{C}_{\tilde{A}^{[]}}S - SC_{\hat{A}^{[*]}} = \mathrm{diag}(I,0,\dots,0), \tag{5.26}$$

then the polynomials

$$\Phi_1(\lambda) = \sum_{k=0}^{n-1} \lambda^k \alpha_n^{-1} s_{n-1,k}, \quad \Psi_1(\lambda) = -\sum_{k=0}^{n-1} \lambda^k s_{k,n-1}\alpha_n^{*-1}$$

satisfy the equation

$$A^{[]}(\lambda)\Phi_1(\lambda) + \Psi_1(\lambda)A^{[*]} = I.$$

Here α_n denotes the leading coefficient of $A^{[]}(\lambda)$ and $\tilde{A}^{[]}(\lambda) = A^{[]}(\lambda)\alpha_n^{-1}$, $\hat{A}^{[*]}(\lambda) = \alpha_n^{*-1}A^{[*]}(\lambda)$. The polynomial

$$\Phi(\lambda) := \tfrac{1}{2}\sum_{k=0}^{n-1} \lambda^k \left(\alpha_n^{-1} s_{n-1,k} - s_{k,n-1}\alpha_n^{*-1}\right)$$

clearly satisfies equation (5.23). Note that in [21] one can find various transformations of the equation (5.26) to a matrix-vector equation $Qx = \rho$, where $\rho \in \mathbb{C}^{nr^2}$ and Q is an $nr^2 \times nr^2$ Hankel or Bezout matrix.

Passing to Step 2, we first note that the resultant matrix $R_{2n-1}(A^{[*]},\Phi)$ is invertible. Indeed, we know from the proof of Theorem 5.1 that there are polynomials $Z(\lambda) = \sum_{j=0}^{n} \lambda^j z_j$ ($\det z_n \neq 0$) and $\Theta(\lambda) = \sum_{j=0}^{n-1} \lambda^j \xi_j$ such that (5.24) holds true. Hence Proposition 3.2 yields that $q_0(A^{[*]},\Phi) \leqslant 2n-1$, i.e.,

$$\mathrm{Ker} R_{2n-1}(A^{[*]},\Phi) = \mathrm{Im\,col}\left(\Phi_F J_F^{k-1}\right)_{k=1}^{2n-1} \dot{+} \mathrm{Im\,col}\left(\Phi_\infty J_\infty^{2n-1-k}\right)_{k=1}^{2n-1},$$

where (Φ_F, J_F) (resp., (Φ_∞, J_∞)) denotes the greatest common restriction of the right finite (resp., infinite) Jordan pairs of $A^{[*]}(\lambda)$ and $\Phi(\lambda)$. We know from (5.23) that the polynomials $A^{[*]}$ and $\Phi(\lambda)$ are right coprime, which means that their right finite Jordan pairs do not have any common restriction. The same is true for the infinite right Jordan pairs of $A^{[*]}$ and $\Phi(\lambda)$, because $A^{[*]}(\lambda)$ has an invertible leading coefficient. So, $\mathrm{Ker} R_{2n-1}(A^{[*]},\Phi) = (0)$, i.e., the square matrix $R_{2n-1}(A^{[*]},\Phi)$ is invertible.

Now we claim that the coefficients of the polynomials $Z(\lambda)$ and $\Theta(\lambda)$ which have the properties stated in Step 2 can bc obtained in the following way. Set

$$z_n = \alpha_n^*, \quad \xi_{n-1} = \alpha_n^{-1}\phi_{n-1}^*\alpha_n^*, \tag{5.27}$$

and solve the equation

$$[R_{2n-1}(A^{[*]},\Phi)]^* \begin{bmatrix} -\xi_0 \\ \vdots \\ -\xi_{n-2} \\ z_0 \\ z_1 \\ \vdots \\ z_{n-1} \end{bmatrix} = \begin{bmatrix} 0 \\ \vdots \\ 0 \\ \alpha_0\xi_{n-1} \\ \alpha_1\xi_{n-1} - \phi_0^* z_n \\ \vdots \\ \alpha_{n-1}\xi_{n-1} - \phi_{n-2}^* z_n \end{bmatrix}, \tag{5.28}$$

where $A^{[*]}(\lambda) = \sum_{j=0}^{n} \lambda^j \alpha_j$ and $\Phi(\lambda) = \sum_{j=0}^{n-1} \lambda^j \phi_j$. Indeed, one easily sees that (5.27)-(5.28) imply (5.24). Moreover, (5.23) and (5.24) yield

$$Z^*(\lambda)\Theta(\lambda) + \Theta^*(\lambda)Z(\lambda) = Z^*(\lambda)[A^{[]}(\lambda)]^{-1}[A^{[*]}(\lambda)]^{-1}Z(\lambda). \tag{5.29}$$

The first equality in (5.27) implies that the rational matrix function $G(\lambda) := [A^{[*]}(\lambda)]^{-1}Z(\lambda)$ is analytic at infinity and $G(\infty) = I$. The same is true for $G^*(\lambda) := Z^*(\lambda)[A^{[]}(\lambda)]^{-1}$. So, the right hand side of (5.29) is analytic at infinity and takes value I there. Since the left hand side of (5.29) is a polynomial, we infer tht (5.25) holds true for any $\lambda \in \mathbb{C}$.

6. MATRIX GENERALIZATION OF THE SECOND M.G. KREIN THEOREM

As already indicated in Section 3, the matrix version of the second Krein theorem deals with a specific inverse problem for block Toeplitz matrices. Namely, given a block column $x = \mathrm{col}(x_j)_{j=0}^{n}$ of r×r matrices ($\det x_0 \neq 0$), find necessary and sufficient conditions in order that there is an invertible hermitian block Toeplitz matrix

$T = [t_{j-k}]_{j,k=0}^{n}$ $(t_j \in \mathbb{C}^{r\times r})$ such that

$$T\,\mathrm{col}(x_j)_{j=0}^{n} = \mathrm{col}(\delta_{0j}I)_{j=0}^{n}. \qquad (6.1)$$

Note that in view of Theorem 3.4, the matrix x_0 is necessarily hermitian. Throughout this section we assume that $x_0 > 0$ and we shall use the notation

$$\mathbf{X}(\lambda) := X^{O}(\lambda)x_0^{-\frac{1}{2}} = \sum_{j=0}^{n} \lambda^j x_{n-j}x_0^{-\frac{1}{2}}.$$

Before stating our main result, we present a simple example which shows that in the block case the polynomial $\mathbf{X}(\lambda)$ may have a pair of eigenvalues that are symmetric with respect to the unit circle. Indeed, take

$$x_0 = \begin{bmatrix} 2 & 0 \\ 0 & \frac{1}{2} \end{bmatrix}, \qquad x_1 = \begin{bmatrix} 1 & 0 \\ 0 & 1 \end{bmatrix}.$$

Then the polynomial $\lambda x_0^{\frac{1}{2}} + x_1 x_0^{-\frac{1}{2}}$ has a symmetric pair of eigenvalues $\lambda_1 = -2$, $\lambda_2 = -\frac{1}{2}$. On the other hand, the block Toeplitz matrix

$$T = \begin{bmatrix} -(I-x_0^2)^{-1}x_0 & (I-x_0^2)^{-1} \\ (I-x_0^2)^{-1} & -(I-x_0^2)^{-1}x_0 \end{bmatrix}$$

is hermitian and invertible and

$$T\begin{bmatrix} x_1 \\ x_0 \end{bmatrix} = \begin{bmatrix} I \\ 0 \end{bmatrix}.$$

Our results will show that the conditions on $\mathbf{X}(\lambda)$ must involve not only the location of the eigenvalues but also some geometric properties of orthogonality nature of the Jordan chains of $\mathbf{X}(\lambda)$.

The approach of this section is based on reducing the above mentioned inverse problem to the problem of coprime symmetric factorization. We first prove the following result.

THEOREM 6.1. *Given a block column* $x = \mathrm{col}(x_j)_{j=0}^{n}$, *with* $x_0 > 0$, *there exists an invertible hermitian block Toeplitz matrix* T *that satisfies*

(6.1) if and only if the function

$$F(\lambda):=[A(\lambda)]^*A(\lambda)\qquad (|\lambda|=1), \tag{6.2}$$

with $A(\lambda):=\mathbf{X}^*(\lambda)$, *admits a right symmetric factorization of order* n

$$F(\lambda):=B(\lambda)[B(\lambda)]^*\qquad (|\lambda|=1) \tag{6.3}$$

that is relatively coprime with (6.2).

If the latter condition holds true and $B(\lambda)=\sum_{j=0}^{n}\lambda^j b_j$ $(b_n>0)$ *is a polynomial that appears in (6.3), then a desired* T *is found as the inverse of the matrix*

$$T^{-1}=\begin{bmatrix} x_0 & & & 0\\ x_1 & \ddots & & \\ \vdots & \ddots & \ddots & \\ x_n & \cdots & x_1 & x_0\end{bmatrix} x_0^{-1}\begin{bmatrix} x_0^* & x_1^* & \cdots & x_n^*\\ & \ddots & \ddots & \vdots\\ & & \ddots & x_1^*\\ 0 & & & x_0^*\end{bmatrix}-\begin{bmatrix} 0 & & & 0\\ b_0 & \ddots & & \\ \vdots & \ddots & \ddots & \\ b_{n-1} & \cdots & b_0 & 0\end{bmatrix}\begin{bmatrix} 0 & b_0^* & \cdots & b_{n-1}^*\\ & \ddots & \ddots & \vdots\\ & & \ddots & b_0^*\\ 0 & & & 0\end{bmatrix} \tag{6.4}$$

Conversely, given an invertible hermitian block Toeplitz matrix T *that satisfies (6.1), the polynomial* $B(\lambda)=\sum_{j=0}^{n}\lambda^j b_j$ *in (6.3) is obtained by setting* $b_j=y_n^{-\frac{1}{2}}y_j$ $(j=0,1,\ldots,n)$, *where* $col(y_j)_{j=0}^{n}$ *is the solution of the equation*

$$T\,col(y_j)_{j=0}^{n}=col(\delta_{nj}I)_{j=0}^{n}, \tag{6.5}$$

and substituting these coefficients b_j *into (6.4) we obtain the inverse of the given matrix* T.

PROOF. To prove the necessity assume that $T=[t_{j-k}]_{j,k=0}^{n}$ is an invertible hermitian block Toeplitz matrix satisfying (6.1) and let $y=col(y_j)_{j=0}^{n}$ be defined by (6.5). We claim that y_n is positive definite. Indeed, Theorem 3.4 yields that y_n is hermitian. Further, introduce the matrices

$$P := \begin{bmatrix} x_0 & & & & \\ x_1 & I & & & \\ x_2 & 0 & \ddots & & \\ \vdots & \vdots & & \ddots & \\ x_n & 0 & \cdots & 0 & I \end{bmatrix}, \qquad Q = \begin{bmatrix} I & 0 & \cdots & 0 & y_0 \\ & \ddots & & \vdots & \vdots \\ & & \ddots & 0 & y_{n-2} \\ & & & I & y_{n-1} \\ & & & & y_n \end{bmatrix}.$$

One easily checks that (cf. [10]),

$$TP = \begin{bmatrix} I & \mathrm{row}(t_{-j})_{j=1}^{n} \\ 0 & T_{n-1} \end{bmatrix},$$

where $T_{n-1} = [t_{j-k}]_{j,k=0}^{n-1}$, and hence the invertibility of x_0 implies that the matrix T_{n-1} is invertible. This implies, in view of the equality

$$TQ = \begin{bmatrix} T_{n-1} & 0 \\ \mathrm{row}(t_{n-j})_{j=0}^{n-1} & I \end{bmatrix},$$

that the matrix Q, and hence y_n, is invertible. One easily computes further (cf. [10]) that

$$P^*TP = \begin{bmatrix} x_0 & 0 \\ 0 & T_{n-1} \end{bmatrix}, \qquad Q^*TQ = \begin{bmatrix} T_{n-1} & 0 \\ 0 & y_n^* \end{bmatrix}, \tag{6.6}$$

which implies that x_0 and $y_n^* = y_n$ have the same inertia. So, the positive definiteness of x_0 yields the same for y_n. Now introduce the polynomial $B(\lambda) = \mathbf{Y}(\lambda) := y_n^{-\frac{1}{2}} \sum_{j=0}^{n} \lambda^j y_j$. Then, in view of Theorem 3.4,

$$B(\lambda)B^{\circledast}(\lambda) = A^{\circledast}(\lambda)A(\lambda) \quad (A(\lambda) := \mathbf{X}^*(\lambda)),$$

and the polynomials $B^{\circledast}(\lambda)$ and $A(\lambda)$ are right coprime.

To prove the sufficient part of the theorem, assume that $B(\lambda) = \sum_{j=0}^{n} \lambda^j b_j$ is a polynomial, with $b_n > 0$, which appears in the right symmetric factorization (6.3). Setting $y_n = b_n^2$ and $Y(\lambda) := B(\lambda)b_n$ we can rewrite (6.2)-(6.3) as

$$Y(\lambda)y_n^{-1}Y^{\circledast}(\lambda) = X(\lambda)x_0^{-1}X^{\circledast}(\lambda).$$

As the factorizations (6.2) and (6.3) are relatively coprime, the polynomials $Y^{\circledast}(\lambda)$ and $X^{\circledast}(\lambda)$ are right coprime. Then Theorem 3.4 ensures the existence of a matrix T with desired properties as well as implies the formula (6.4).

□

The above result, in conjunction with Theorem 5.1, leads to the following matrix generalization of the second M.G. Krein theorem.

THEOREM 6.2. *Let* $x = \mathrm{col}(x_j)_{j=0}^{n}$ *be a given block column with positive definite* x_0. *There exists an invertible hermitian block Toeplitz matrix T satisfying (6.1) if and only if for every symmetric pair of eigenvalues* $\lambda_0, \bar{\lambda}_0^{-1}$ *of* $\mathbf{X}(\lambda)$ *(if any), the equalities*

$$\sum_{i=0}^{k} \langle \phi_i, \psi_{k-i} \rangle = 0 \quad \left(k=0,1,\ldots,\min\{\alpha,\beta\}\right) \tag{6.7}$$

hold true for any left Jordan chains $\phi_0,\ldots,\phi_\alpha$ *and* $\psi_0,\ldots,\psi_\beta$ *of* $\mathbf{X}(\lambda)$, *corresponding to* λ_0 *and* $\bar{\lambda}_0^{-1}$, *respectively.*

It is clear that in the scalar case, Theorem 6.2 becomes Krein's second theorem. Moreover, if the (scalar) polynomial $\mathbf{X}(\lambda)$ does not have symmetric pairs of zeroes, there is a unique choice of the polynomial $B(\lambda)$ in (6.3), namely, $B(\lambda) = \mathbf{X}^*(\lambda)$. Consequently in this case, in view of Theorem 6.1, there is a unique invertible hermitian Toeplitz matrix satisfying (6.1), which is given as the inverse of the following matrix (cf.[8],[17])

$$T^{-1} = x_0^{-1}\left\{\begin{bmatrix} x_0 & & & 0 \\ x_1 & \ddots & & \\ \vdots & \ddots & \ddots & \\ x_n & \cdots & x_1 & x_0 \end{bmatrix}\begin{bmatrix} \bar{x}_0 & \bar{x}_1 & \cdots & \bar{x}_n \\ & \ddots & \ddots & \vdots \\ & & \ddots & \bar{x}_1 \\ 0 & & & \bar{x}_0 \end{bmatrix} - \begin{bmatrix} 0 & & & 0 \\ \bar{x}_n & \ddots & & \\ \vdots & \ddots & \ddots & \\ \bar{x}_1 & \cdots & \bar{x}_n & 0 \end{bmatrix}\begin{bmatrix} 0 & x_n & \cdots & x_1 \\ & \ddots & \ddots & \vdots \\ & & \ddots & x_n \\ 0 & & & 0 \end{bmatrix}\right\} \tag{6.8}$$

Consider now the case where the matrix T is required to be

positive definite. The next result is a matrix generalization of the well known G. Szegö theorem on orthogonal polynomials (see [27],[9]).

COROLLARY 6.3. *Given a block column* $x = \mathrm{col}(x_j)_{j=0}^n$, *there exists a positive definite block Toeplitz matrix* T *satisfying (6.1) if and only if* $x_0 > 0$ *and the function* $\det X(z)$ *has no zeroes outside the unit circle.*

PROOF. The "if" part follows directly from Theorem 6.2. To prove the "only if" part, first note that in view of (6.6) $x_0 > 0$ and $T_{n-1} > 0$. Then applying Theorem 2.1 we conclude that all zeroes of $\det X(z)$ are inside the unit circle.

□

Note that matrix generalizations of G. Szegö's results were known prior to this (see e.g., [7]).

7. INVERSE PROBLEMS FOR GENERAL BLOCK TOEPLITZ MATRICES AND STEIN EQUATIONS

In this section we start with a general inverse problem of determining a $(n+1)r \times (n+1)r$ block Toeplitz matrix $T = [t_{j-k}]_{j,k=0}^n$ such that

$$T\,\mathrm{col}(x_j)_{j=0}^n = \mathrm{col}(\delta_{0j}I)_{j=0}^n, \qquad \mathrm{row}(v_j)_{j=0}^n T = \mathrm{row}(\delta_{0j}I)_{j=0}^n. \tag{7.1}$$

where $x = \mathrm{col}(x_j)_{j=0}^n$ and $v = \mathrm{row}(v_j)_{j=0}^n$ are given block vectors with $\det x_0 \neq 0 \neq \det v_0$. Note that this problem differs from the type of inverse problems discussed in Section 3, where the matrix T is additionally required to be invertible. Specifying the solution of the above stated problem for the case of positive definite x_0 and hermitian T, we shall obtain in the next section a simple proof of Theorem 6.2, along with efficient formulas for the desired matrix T and for the factors in a right symmetric factorization that is relatively coprime with a given left one.

The solutions of the problems under consideration will be expressed in terms of solutions of the following linear matrix equation of Stein type:

$$S - \hat{K}_{\hat{V}^O} S K_{\tilde{X}^O} = \mathrm{diag}(x_0^{-1}, 0, \ldots, 0), \tag{7.2}$$

where $\hat{K}_A$ and K_A are defined by (2.2) and

$$\hat{V}^O(\lambda) := \sum_{j=0}^{n} \lambda^j v_0^{-1} v_{n-j}, \quad \tilde{X}^O := \sum_{j=0}^{n} \lambda^j x_{n-j} x_0^{-1}. \tag{7.3}$$

THEOREM 7.1. *Given block vectors* $x = \mathrm{col}(x_j)_{j=0}^{n}$ $(\det x_0 \neq 0)$ *and* $v = \mathrm{row}(v_j)_{j=0}^{n}$, *there exists a block Toeplitz matrix* T *that satisfies equations (7.1) if and only if* $v_0 = x_0$ *and one of the following equivalent conditions is fulfilled:*

(a) the matrix equation (7.2) is solvable;

(b) there are matrix polynomials $\Phi(\lambda)$ *and* $\Psi(\lambda)$ *that satisfy the equation*

$$V^O(\lambda)\Phi(\lambda) + \Psi(\lambda)X(\lambda) = x_0; \tag{7.4}$$

(c) if λ_0 *is a common eigenvalue of the polynomials* $X(\lambda)$ *and* $V^O(\lambda)$, *then for any right Jordan chain* $g_0, g_1, \ldots, g_\beta$ *of* $X(\lambda)$ *and for any left Jordan chain* $f_0, f_1, \ldots, f_\alpha$ *of* $V^O(\lambda)$, *corresponding to* λ_0, *the equalities*

$$\sum_{j=0}^{k} f_j x_0 g_{n-j} = 0 \quad (k=0,1,\ldots,\max\{\alpha,\beta\})$$

hold true.

If the above conditions are fulfilled, then any solution S *of (7.2) generates a block Toeplitz matrix* T *satisfying (7.1) by the formula*

$$T = \left[\begin{array}{c|cccc} v_0^{-1} & -v_0^{-1}v_1 & \cdots & & -v_0^{-1}v_n \\ \hline & I & & & \\ & & \ddots & & \\ 0 & & & \ddots & \\ & & & & I \end{array}\right] \left[\begin{array}{c|c} x_0 & 0 \\ \hline 0 & S \end{array}\right] \left[\begin{array}{c|cccc} x_0^{-1} & & & 0 & \\ \hline -x_1x_0^{-1} & I & & & \\ \vdots & & \ddots & & \\ & & & \ddots & \\ -x_nx_0^{-1} & & & & I \end{array}\right]. \tag{7.5}$$

Conversely, given a block Toeplitz matrix $T = [t_{j-k}]_{j,k=0}^{n}$ *that satisfies (7.1), the matrix* $S = T_{n-1} = [t_{j-k}]_{j,k=0}^{n-1}$ *is a solution of the equation (7.2), and substituting this solution* S *into (7.5) one obtains the given* T.

PROOF. The equivalence of conditions *(a)*, *(b)* and *(c)* follows from Theorem 4.5. Thus we can deal with condition *(a)* only. To prove the necessary part of the theorem, assume that $T = [t_{j-k}]_{j,k=0}^{n}$ is a block Toeplitz matrix satisfying (7.1). Multiplying both sides of the first (second) equation in (7.1) by $\text{row}(v_j)_{j=0}^{n}$ $\left(\text{col}(x_j)_{j=0}^{n}\right)$, from the left (right) we obtain

$$v_0 = [v_0 v_1 \dots v_n] T \begin{bmatrix} x_0 \\ x_1 \\ \vdots \\ x_n \end{bmatrix} = x_0$$

Furthermore, in view of Step 1 in the proof of Theorem 2.1, the equation (7.2) has a solution $S = [t_{j-k}]_{j,k=0}^{n-1}$, and the necessary part follows.

To prove the sufficient part of the theorem, assume that $x_0 = v_0$ and equation (7.2) is solvable. Let S be a solution of (7.2). Then S must be a block Toeplitz matrix (see e.g., [21]), say $S = [s_{j-k}]_{j,k=0}^{n-1}$. We shall show that the matrix T defined by (7.5) is block Toeplitz as well. Introduce the notations

$$\gamma = \begin{bmatrix} -x_1 x_0^{-1} \\ \vdots \\ -x_n x_0^{-1} \end{bmatrix}, \qquad \delta = [-v_0^{-1} v_1 \dots -v_0^{-1} v_n]$$

and rewrite (7.5) as follows

$$T = \left[\begin{array}{c|c} v_0^{-1} + \delta S \gamma & \delta S \\ \hline S\gamma & S \end{array}\right] \tag{7.6}$$

We have to show that

$$v_0^{-1} + \delta S \gamma = s_0 \tag{7.7}$$

and

$$\delta S = [s_{-1} \dots s_{-n+1} \ *], \quad S\gamma = \begin{bmatrix} s_1 \\ \vdots \\ s_{n-1} \\ * \end{bmatrix}, \tag{7.8}$$

where $*$ denotes unspecified entries. To this end note that

$$\hat{K}_{\hat{V}0} = \begin{bmatrix} \delta \\ \Delta \end{bmatrix}, \quad K_{\tilde{X}0} = \begin{bmatrix} \gamma & \Delta^{\tau} \end{bmatrix},$$

where

$$\Delta = \underbrace{\begin{bmatrix} I & 0 & \cdots & 0 \\ & \ddots & \ddots & \vdots \\ & & \ddots & \vdots \\ 0 & & I & 0 \end{bmatrix}}_{n} \Big\} n-1 ,$$

and rewrite equation (7.2) in the form

$$S - \begin{bmatrix} \delta S\gamma & \delta S\Delta^{\tau} \\ \Delta S\gamma & \Delta S\Delta^{\tau} \end{bmatrix} = \operatorname{diag}\left(x_0^{-1}, 0, \dots, 0\right). \tag{7.9}$$

Comparing the entries on the right and on the left in (7.9), we obtain

$$s_0 - \delta S\gamma = x_0^{-1}, \tag{7.10}$$

$$[s_{-1} \dots s_{-n+1}] - \delta S\Delta^{\tau} = 0 , \quad \begin{bmatrix} s_1 \\ \vdots \\ s_{n-1} \end{bmatrix} - \Delta S\gamma = 0. \tag{7.11}$$

It is clear that (7.10) is the same as (7.7), and (7.11) implies (7.8), which proves that the matrix T defined by (7.5) is block Toeplitz. Also, one easily checks that such a matrix T satisfies equations (7.1), which completes the proof.

□

Note that, under the assumption $v_0 = x_0$, the fact that the existence of a block Toeplitz matrix satisfying (7.1) is equivalent to the solvability of a Stein-type matrix equation (which is somewhat different from (7.2)), was stated (without proof) in [21].

Now consider the problem of determining a *hermitian* block Toeplitz matrix that satisfies the equation

$$T\ \mathrm{col}\left(x_j\right)_{j=0}^{n} = \mathrm{col}\left(\delta_{0j} I\right)_{j=0}^{n}, \tag{7.12}$$

where $x = \mathrm{col}\left(x_j\right)_{j=0}^{n}$ is given and x_0 is hermitian and invertible. To put this problem into the framework of Theorem 7.1, we have to set $v_j = x_j^*$ and replace equation (7.2) by the equation

$$S - \left(K_{\tilde{X}0}\right)^* S K_{\tilde{X}0} = \mathrm{diag}\left(x_0^{-1}, 0, \ldots, 0\right). \tag{7.13}$$

Observe that this equation is solvable if and only if it has a hermitian solution. Also note that $\phi_0, \ldots, \phi_\beta$ is a left Jordan chain of $X^{\circledast}(\lambda)$ corresponding to λ_0 if and only if $\bar{\phi}_0, \ldots, \bar{\phi}_\beta$ is a right Jordan chain of $X(\lambda)$ corresponding to $\bar{\lambda}_0^{-1}$. With these remarks, Theorem 7.1 leads to the following result.

THEOREM 7.2. *Given a block vector* $x = \left(x_j\right)_{j=0}^{n}$, *with invertible* x_0, *there exists a hermitian block Toeplitz matrix* T *that satisfies equation (7.12) if and only if* x_0 *is hermitian and one of the following equivalent conditions is fulfilled:*

(a') the matrix equation (7.13) is solvable;

(b') there are matrix polynomials $\Phi(\lambda)$ *and* $\Psi(\lambda)$ *that satisfy the equation*

$$X^{\circledast}(\lambda)\Phi(\lambda) + \Psi(\lambda)X(\lambda) = x_0; \tag{7.14}$$

(c') for every symmetric pair of eigenvalues $\lambda_0, \bar{\lambda}_0^{-1}$ *of* $X(\lambda)$ *(if any), the equalities*

$$\sum_{j=0}^{k} \langle x_0 g_{k-j}, f_j \rangle = 0 \quad (k=0,1,\ldots,\min\{\alpha,\beta\}) \tag{7.15}$$

hold true for any right Jordan chains $f_0,\ldots,f_\alpha$ *and* $g_0,\ldots,g_\beta$ *of* $X(\lambda)$ *corresponding to* λ_0 *and* $\bar{\lambda}_0^{-1}$, *respectively.*

If the above conditions are fulfilled, then any hermitian solution S *of (7.13) generates a hermitian block Toeplitz matrix* T *satisfying (7.12) by the formula*

$$T = \left[\begin{array}{c|ccc} x_0^{-1} & -x_0^{-1}x_1^* & \cdots & -x_0^{-1}x_n^* \\ \hline & I & & \\ 0 & & \ddots & \\ & & & I \end{array}\right] \left[\begin{array}{c|c} x_0 & 0 \\ \hline 0 & S \end{array}\right] \left[\begin{array}{c|ccc} x_0^{-1} & & 0 & \\ \hline -x_1x_0^{-1} & I & & \\ \vdots & & \ddots & \\ -x_nx_0^{-1} & & & I \end{array}\right] \tag{7.16}$$

Conversely, given a hermitian block Toeplitz matrix $T = [t_{j-k}]_{j,k=0}^{n}$ *that satisfies (7.12), the matrix* $S = T_{n-1} = [t_{j-k}]_{j,k=0}^{n-1}$ *is a hermitian solution of the equation (7.13), and substituting this solution* S *into (7.16) one obtains the given* T.

Another interesting approach to the problem under consideration was developed by A. Atzmon in [2]. By a different method he proved that the existence of a hermitian block-Toeplitz matrix T satisfying (7.12) is equivalent to the solvability of the equation (7.14), and he obtained a formula for T via the solutions of (7.14).

8. INVERSE PROBLEMS FOR INVERTIBLE HERMITIAN BLOCK TOEPLITZ MATRICES

In this section we return to the problem of determining an *invertible* hermitian block Toeplitz matrix T given the first block column

$x = \mathrm{col}(x_j)_{j=0}^{n}$ of its inverse T^{-1}. Theorem 7.2 and, in particular formula (7.16), suggest that such a matrix T exists if and only if there is an invertible hermitian solution of the matrix equation (7.13). The result which follows is based on the fact that if $x_0 > 0$ and (7.13) is solvable, then any hermitian solution S of (7.13) is invertible.

THEOREM 8.1. *Let* $x = \mathrm{col}(x_j)_{j=0}^{n}$ *be a given block column with positive definite* x_0 *and denote* $\mathbf{X}(\lambda) := \sum_{j=0}^{n} \lambda^n x_{n-j} x_0^{-1/2}$. *There exists an invertible hermitian block Toeplitz matrix* T *such that*

$$T\,\mathrm{col}(x_j)_{j=0}^{n} = \mathrm{col}(\delta_{0j} I)_{j=0}^{n}, \tag{8.1}$$

if and only if one of the following equivalent conditions is fulfilled:

(i) *the matrix equation*

$$S - K_{\tilde{X}}^{*} S K_{\tilde{X}} = \mathrm{diag}(x_0^{-1}, 0, \ldots, 0) \tag{8.2}$$

is solvable;

(ii) *the equation*

$$\mathbf{X}^{\circledast}(\lambda)\Phi(\lambda) + \Psi(\lambda)\mathbf{X}(\lambda) = I \tag{8.3}$$

is solvable in matrix polynomials;

(iii) *for every symmetric pair of eigenvalues* $\lambda_0, \bar{\lambda}_0^{-1}$ *of* $\mathbf{X}(\lambda)$ *(if any) the equalities*

$$\sum_{j=0}^{k} \langle \phi_{k-j}, \psi_j \rangle = 0 \quad (k=0,1,\ldots,\max\{\alpha,\beta\}) \tag{8.4}$$

hold true for any right Jordan chains $\phi_0, \ldots, \phi_\alpha$ *and* $\psi_0, \ldots, \psi_\beta$ *of* $\mathbf{X}(\lambda)$, *corresponding to* λ_0 *and* $\bar{\lambda}_0^{-1}$, *respectively.*

If the above conditions are fulfilled, then any hermitian solution S *of (8.2) generates an invertible hermitian block Toeplitz matrix* T *satisfying (8.1), by the formula (7.16). Conversely, given an invertible hermitian block Toeplitz matrix* $T = [t_{j-k}]_{j,k=0}^{n}$ *that satisfies (8.1), the matrix* $S = T_{n-1} = [t_{j-k}]_{j,k=0}^{n-1}$ *is a hermitian invertible solution of (8.2) and the given matrix* T *is obtained by substituting* $S = T_{n-1}$ *into (7.16).*

Before turning to the proof of Theorem 8.1, we present a simple example which shows that the assumption $x_0 > 0$ in this theorem is essential. Indeed, take $x_0 = \begin{bmatrix} 1 & 0 \\ 0 & -1 \end{bmatrix}$, $x_1 = \begin{bmatrix} 0 & -1 \\ 0 & 0 \end{bmatrix}$. Then $\tilde{\mathcal{X}}(\lambda) = \begin{bmatrix} \lambda & 1 \\ 0 & \lambda \end{bmatrix}$ and equation (8.2) becomes

$$\begin{bmatrix} s_{11} & s_{12} \\ s_{21} & s_{22} \end{bmatrix} - \begin{bmatrix} 0 & 0 \\ 1 & 0 \end{bmatrix} \begin{bmatrix} s_{11} & s_{12} \\ s_{21} & s_{22} \end{bmatrix} \begin{bmatrix} 0 & 1 \\ 0 & 0 \end{bmatrix} = \begin{bmatrix} 1 & 0 \\ 0 & -1 \end{bmatrix} .$$

The last equation has a unique solution $S = \begin{bmatrix} 1 & 0 \\ 0 & 0 \end{bmatrix}$ and, in view of Theorem 7.2, there is a *unique* block Toeplitz matrix T such that $T\begin{bmatrix} x_0 \\ x_1 \end{bmatrix} = \begin{bmatrix} I \\ 0 \end{bmatrix}$, which is given by

$$T = \begin{bmatrix} 1 & 0 & 0 & 0 \\ 0 & 0 & -1 & 0 \\ 0 & -1 & 1 & 0 \\ 0 & 0 & 0 & 0 \end{bmatrix} .$$

This matrix T is clearly singular.

PROOF OF THEOREM 8.1. Firstly, note that $\tilde{X}(\lambda) = \tilde{X}^{O}(\lambda)$, and hence equations (8.2) and (7.13), coincide. Secondly, equation (7.14) can be rewritten as

$$X^*(\lambda)\Psi^*(\lambda) + \Phi^*(\lambda)X^{O}(\lambda) = x_0,$$

and multiplying both sides of this equation by $x_0^{-\frac{1}{2}}$, from the two sides we see that (8.3) is solvable if and only if (7.14) is solvable. Thirdly, a sequence $f_0,\dots,f_\alpha$ is a right Jordan chain of $X(\lambda)$ corresponding to $\lambda_0 \neq 0$ if and only if $x_0^{\frac{1}{2}}f_0,\dots,x_0^{\frac{1}{2}}f_\alpha$ is a right Jordan chain of $\mathcal{X}(\lambda)$ corresponding to λ_0^{-1}, and hence condition *(iii)* coincides with *(c')*. So, we conclude that conditions *(i)*, *(ii)* and *(iii)* coincide with conditions *(a')*, *(b')* and *(c')*, respectively, and hence, in view of Theorem 4.5, they are equivalent one to another. Now in order to prove the theorem, it is enough to show that the assumption $x_0 > 0$ implies that any hermitian solution of (8.2) is invertible. Indeed, if S is a hermitian solution of (8.2), then, as shown in Step 2 of the proof of Theorem 2.1, S satisfies the equation

$$S - (K_{\tilde{X}}^*)^n S K_{\tilde{X}}^n = \Gamma^* \mathrm{diag}(x_0^{-1}, \ldots, x_0^{-1})\Gamma, \tag{8.5}$$

where Γ us defined by (2.17). Condition *(iii)* implies, in particular, that $\det X(\lambda) \neq 0$, $|\lambda|=1$. It follows that the matrix $K_{\tilde{X}}$, and hence $K_{\tilde{X}}^n$, does not have eigenvalues on the unit circle. Then the well known inertia results (see e.g., [18], [19], [26], [28], [29]) lead to the conclusion that S is invertible. Conversely, if $T = [t_{j-k}]_{j,k=0}^n$ satisfies (8.1) and $x_0 > 0$, then a simple argument (see the proof of Theorem 6.1) shows that $T_{n-1} = [t_{j-k}]_{j,k=0}^{n-1}$ is invertible and from Step 1 in the proof of Theorem 2.1 we know that $S = T_{n-1}$ satisfies equation (8.2).

□

Combining this theorem with the results of Section 6, we obtain the following result on coprime symmetric factorizations.

THEOREM 8.2. *Let* $A(\lambda) = \sum_{j=0}^{n} \lambda^j a_j$ *be a rxr matrix polynomial with positive definite leading coefficient* a_n. *Then the function*

$$F(\lambda) := [A(\lambda)]^* A(\lambda) \quad (|\lambda|=1) \tag{8.6}$$

admits a right symmetric factorization of order n

$$F(\lambda) = B(\lambda)[B(\lambda)]^* \quad (|\lambda|=1) \tag{8.7}$$

that is relatively coprime with (8.6), if and only if the Stein equation

$$S - K_{\tilde{A}^*}^* S K_{\tilde{A}^*} = \mathrm{diag}(a_n^{-2}, 0, \ldots, 0) \tag{8.8}$$

is solvable.

If S is a solution of (8.8), then the coefficients of a polynomial $B(\lambda) = \sum_{j=0}^{n} \lambda^j b_j$ *in (8.7) are given by the formulas*

$$b_j = (a_0^* a_0 + c_n)^{-1/2} (a_{n-j}^* a_0 + c_j) \quad (j=0,1,\ldots,n), \tag{8.9}$$

where $c_0 = 0$ *and* $\mathrm{col}(c_j)_{j=1}^n$ *is found from the equation*

$$S\ \mathrm{col}(c_j)_{j=1}^n = (\delta_{nj} I)_{j=1}^n . \qquad (8.10)$$

Conversely, given a factorization (8.7), a solution S *of (8.8) can be found as the inverse of the following matrix*

$$S^{-1} = \begin{bmatrix} a_n^* & & 0 \\ \vdots & \ddots & \\ a_0^* & \cdots & a_n^* \end{bmatrix} \begin{bmatrix} a_n & \cdots & a_0 \\ & \ddots & \vdots \\ 0 & & a_n \end{bmatrix} - \begin{bmatrix} b_0 & & 0 \\ \vdots & \ddots & \\ b_{n-1} & \cdots & b_0 \end{bmatrix} \begin{bmatrix} b_0^* & \cdots & b_{n-1}^* \\ & \ddots & \vdots \\ 0 & & b_0^* \end{bmatrix}, \qquad (8.11)$$

and substituting this solution S *into (8.9)-(8.10) one obtains the given polynomial* $B(\lambda)$.

PROOF. Assume first that the factorization (8.7) exists. Then in view of Theorem 6.1 there is an invertible hermitian block Toeplitz matrix $T = [t_{p-q}]_{p,q=0}^n$ such that (8.1) holds true with

$$x_j = a_{n-j}^* a_n \quad (j=0,1,\dots,n). \qquad (8.12)$$

It follows from Theorem 8.1 that the equation (8.2), which coincides with (8.8), is solvable. Conversely, assume that (8.8) is solvable. Then in view of Theorem 8.1 there is an invertible hermitian block Toeplitz matrix T that satisfies (8.1), with x_0 defined by (8.12). Applying again Theorem 6.1, we conclude that the factorization (8.7) exists.

Now let S be a solution of (8.8), and let T be the invertible hermitian block Toeplitz matrix given by (7.16), with x_j defined by (8.12). We know from Theorem 6.1 that the polynomial $B(\lambda) = \sum_{j=0}^n \lambda^j b_j$ in (8.7) is obtained by setting $b_j = y_n^{-1/2} y_j$, $(j=0,1,\dots,n)$, where $\mathrm{col}(y_j)_{j=0}^n = T^{-1}\mathrm{col}(\delta_{nj} I)_{j=0}^n$. From (7.16) we have

$$T^{-1} = \left[\begin{array}{c|ccc} x_0 & & 0 & \\ \hline x_1 & I & & \\ \vdots & & \ddots & \\ x_n & & & I \end{array}\right] \left[\begin{array}{c|c} x_0^{-1} & 0 \\ \hline 0 & S^{-1} \end{array}\right] \left[\begin{array}{c|ccc} x_0^* & x_1^* & \cdots & x_n^* \\ \hline & I & & \\ 0 & & \ddots & \\ & & & I \end{array}\right], \qquad (8.13)$$

which yields (8.9)-(8.10).

Conversely, let $B(\lambda) = \sum_{j=0}^{n} \lambda^j b_j$ be given. Then in view of Theorem 6.1, formula (6.4) gives the inverse of an invertible hermitian block Toeplitz matrix $T = [t_{p-q}]_{p,q=0}^{n}$ that satisfies (8.1), with x_j defined by (8.12). Moreover, in this case $b_j = y_n^{-½} y_j$ $(j=0,1,\dots,n)$, where

$$T\ \mathrm{col}\left(y_j\right)_{j=0}^{n} = \mathrm{col}\left(\delta_{nj} I\right)_{j=0}^{n}. \tag{8.14}$$

On the other hand, we know from Theorem 8.1 that the matrix $S = T_{n-1} = [t_{p-q}]_{p,q=0}^{n-1}$ is a solution of the equation (8.8), and substituting this solution S into the right hand side of (8.13) we obtain T^{-1}. Hence setting $S = T_{n-1}$ in (8.9)-(8.10) we obtain the given polynomial $B(\lambda)$. Since T satisfies (8.1) and (8.14) we obtain (8.11) from (3.16') by substituting $x_j = a_{n-j}^* a_n$, $y_j = b_n b_j$ $(j=0,1,\dots,n)$. □

Consider now the case when $\det A(z)$ has all zeroes inside the unit circle (cf. Corollary 6.3). In this case $\lambda\bar{\mu} \neq 1$ for any $\lambda,\mu \in \sigma(\tilde{A}^*)$, and hence equation (8.8) has a unique solution S. It follows from Theorem 8.2 that the polynomial $B(\lambda)$ in (8.7) is uniquely determined. On the other hand, well known results (see [5], [14]) on Wiener-Hopf factorization imply that the function $F(\lambda)$ defined by (8.6) admits a right canonical factorization of the form

$$F(\lambda) = R(\lambda)[R(\lambda)]^* \quad (|\lambda|=1), \tag{8.15}$$

with

$$R(\lambda) = r_n + \lambda^{-1} r_{n-1} + \dots + \lambda^{-n} r_n \tag{8.16}$$

where $r_n > 0$ and $R(\lambda)$ is invertible outside the unit circle. Clearly, setting $B(\lambda) := \lambda^n R(\lambda)$ we obtain the right symmetric factorization (8.7). As a result of the above consideration we have the following result.

COROLLARY 8.3. *Let $A(\lambda)$ be as in Theorem 8.2 and let $\det A(z)$ have no zeroes outside the unit circle. Then the right symmetric factorization (8.7) which is relatively coprime with (8.6) always exists and is unique. The polynomial $B(\lambda)$ in (8.7) has no eigenvalues outside the unit circle and can be obtained from the right Wiener-Hopf factorization (8.15)-(8.16) by setting $B(\lambda) = \lambda^n R(\lambda)$.*

Observe that in the case under consideration, equation (8.6) can be rewritten as a left Wiener-Hopf factorization

$$F(\lambda) = [L(\lambda)]^* L(\lambda) \qquad (|\lambda|=1), \tag{8.17}$$

by setting $L(\lambda) = \lambda^{-n} A(\lambda)$. So, given a left Wiener-Hopf canonical factorization (8.17) of $F(\lambda)$, formulas (8.9)-(8.10) allow the factors to be expressed in the right Wiener-Hopf factorization (8.15) of $F(\lambda)$. For the sake of completeness this result is explicitly stated below.

COROLLARY 8.4. *Given a left canonical Wiener-Hopf factorization*

$$F(\lambda) := [L(\lambda)]^* L(\lambda) \qquad (|\lambda|=1),$$

where $L(\lambda) = \sum_{j=0}^{n} \lambda^{-j} \ell_{n-j}$, $\ell_n > 0$, and $L(\lambda)$ is invertible outside the unit circle. Then a right canonical Wiener-Hopf factorization of $F(\lambda)$,

$$F(\lambda) = R(\lambda)[R(\lambda)]^* \quad (|\lambda|=1),$$

$\det R(\lambda) \neq 0$ $(|\lambda|>1)$, is obtained by setting

$$R(\lambda) = r_n + \lambda^{-1} r_{n-1} + \dots + \lambda^{-n} r_0,$$

where the coefficients $r_0, r_1, \dots, r_n$ are found as follows:

Let S be the unique solution of the Stein equation

$$S - \Lambda S \Lambda^* = diag(\ell_n^{-2}, 0, \dots, 0),$$

where

$$\Lambda := \begin{bmatrix} -\ell_n^{-1}\ell_{n-1} & -\ell_n^{-1}\ell_{n-2} & \cdots & -\ell_n^{-1}\ell_0 \\ I & & & 0 \\ & \ddots & & \vdots \\ & & I & 0 \end{bmatrix},$$

and set $\operatorname{col}(c_j)_{j=1}^{n} = S^{-1}\operatorname{col}(\delta_{nj}I)_{j=1}^{n}$, $c_0 = 0$. *Then*

$$r_j = (\ell_0^*\ell_0+c_n)^{-\frac{1}{2}}(\ell_{n-j}\ell_0+c_j) \quad (j=0,1,\ldots,n).$$

In conclusion we recall that the considerations at the end of Section 5 provide another algorithm of determining a right canonical Wiener-Hopf factorization of $F(\lambda)$, given its left one (8.17).

REFERENCES

[1] Alpay, D. and Gohberg, I.: On orthogonal matrix polynomials, Operator Theory: Advances and Applications, this issue.

[2] Atzmon, A.: n-orthonormal operator polynomials, Operator Theory: Advances and Applications, this issue.

[3] Ball, J.A. and Ran, A.C.M.: Local inverse problems for rational matrix functions, Integral Equations and Operator Theory, 10 (1987), 349-415.

[4] Ben-Artzi, A. and Gohberg, I: Extension of a theorem of Krein on orthogonal polynomials for the non-stationary case, this issue.

[5] Clancey, K. and Gohberg, I.: Factorization of matrix functions and singular integral operators. Operator Theory: Advances and Applications, Vol.3., Birkhäuser Verlag, Basel, 1981.

[6] Daleckii, Iu. L. and Krein, M.G.: Stability of solutions of differential equations in Banach space. Amer. Math. Soc. Transl. 43, American Mathematical Society, Providence, R.I., 1974.

[7] Delsarte, P., Genin, Y.V. and Kamp, Y.G.: Orthogonal polynomial matrices on the unit circle, IEEE Trans. Circuits Syst., CAS-25(3) (1978), 149-160.

[8] Ellis, R.L., Gohberg, I. and Lay, D.C.: On two theorems of M.G. Krein concerning polynomials orthogonal on the unit circle, Integral Equations and Operator Theory, 11 (1988), 87-104.

[9] Geronimus, Ya. L.: Polynomial orthogonal on a circle and interval, Pergamon Press, 1960, (translated from Russian).

[10] Gohberg, I.C. and Heinig, G.: Inversion of finite section Toeplitz matrices consisting of elements of a non-commutative algebra, Rev. Roum. Math. Pures et Appl., 19 (5) (1974), 623-663 (Russian).

[11] Gohberg, I., Kaashoek, M.A., Lerer, L. and Rodman, L.: Common multiples and common divisors of matrix polynomials, I: Spectral method, Indiana Univ. Math. J. 30 (1981), 321-356.

[12] Gohberg, I., Kaashoek, M.A., Lerer, L., and Rodman, L: Common multiples and common divisors of matrix polynomials, II: Vandermonde and resultant, Linear and Multinlinear Algebra 12 (1982), 159-203.

[13] Gohberg, I., Lancaster, P. and Rodman, L.: Matrix Polynomials, Academic Press, New York, 1982.

[14] Gohberg, I., Lancaster, P. and Rodman, L.: Matrices and indefinite scalar products. Operator theory: Advances and Applications, Vol. 8, Birkhäuser Verlag, Basel, 1983.

[15] Gohberg, I. and Lerer, L.: Resultants of matrix polynomials, Bull. Amer. Math. Soc. 82 (1976), No.4, 565-567.

[16] Gohberg, I. and Lerer, L.: On solution of the equation $A(\lambda)X(\lambda) + Y(\lambda)B(\lambda) = C(\lambda)$ in matrix polynomials, unpublished manuscript.

[17] Gohberg, I.C., and Semenčul, A.A.: On the inversion of finite Toeplitz matrices and their continuous analogues, Math. Issled, 7 (2) (1972), 272-283 (Russian).

[18] Hill, R.D.: Inertia theory for simultaneously triangulable complex matrices, Linear Algebra Appl. 2 (1969), 131-142.

[19] Krein, M.G.: Stability theory of differential equations in Banach spaces, Kiev, 1964 (Russian, an expanded version of this book is [6]).

[20] Krein, M.G.: Distribution of roots of polynomials orthogonal on the unit circle with respect to a sign alternating weight, Theor. Funkcii Functional Anal. i Prilozen. 2 (1966), 131-137 (Russian).

[21] Lancaster, P., Lerer, L. and Tismenetsky, M.: Factored form of solutions of the equation AX-XB = C in matrices, Linear Algebra Appl., 62 (1984), 19-49.

[22] Lancaster, P. and Tismenetsky, M.: The Theory of Matrices, Academic Press, Orlando, 1985.

[23] Lerer, L. and Tismenetsky, M.: The eigenvalue separation problem for matrix polynomials, Integral Equations Operator Theory 5, (1982), 386-445.

[24] Lerer, L. and Tismentsky, M.: Bezoutian for several matrix polynomial and matrix equations, Technical Report 88.145, IBM-Israel Scientific Center, Haifa, November 1984.

[25] Lerer, L. and Tismentsky, M.: Generalized bezoutian and matrix equations, Linear Algebra Appl., in press.

[26] Ostrowski, A. and Schneider H.: Some theorems on the inertia of general matrices, J. Math. Anal. Appl., 4 (1962), 72-84.

[27] Szego, G.; Orthogonal polynomials, Colloquium Publications, No.23, American Mathematical Society, Providence, R.I. 2nd ed. 1958, 3rd ed. 1967.

[28] Taussky, O.: Matrices C with $C^n \to 0$, J. Algebra 1 (1969), 5-10.

[29] Wimmer, H.: On the Ostrowski-Schneider inertia theorem, J. Math. Anal. Appl. 41 (1973), 164-173.

I. Gohberg
School of Mathematical Sciences
Raymond and Beverly Sackler
Faculty of Exact Sciences
Tel Aviv University
Tel Aviv 69978 Israel

L. Lerer
Department of Mathematics
Technion—Israel Institute of Technology
Haifa 32000 Israel

Operator Theory:
Advances and Applications, Vol. 34

POLYNOMIALS ORTHOGONAL IN AN INDEFINITE METRIC

H. J. Landau

Introduction

A nonnegative measure $d\mu$ on the unit circle $|z| = 1$ generates a scalar product $[\cdot, \cdot]$ by the formula

$$[P, S] \equiv \int_{|z|=1} P(z)\overline{S(z)}\, d\mu . \qquad (1)$$

On applying the Gram-Schmidt process to orthogonalize the successive powers of z, we obtain *orthogonal polynomials* $1 = P_0(z),\ P_1(x), \ldots$ defined by the requirement that, for each $k \geq 1$,

$$[S_{k-1}, P_k] = 0 ,$$

with S_{k-1} any polynomial of degree less than k. For polynomials of degree n, the scalar product (1) can be calculated explicitly from the coefficients using the first n moments of $d\mu$,

$$c_k \equiv \int_{|z|=1} z^k\, d\mu ,$$

by the expression

$$[P, S] = \sum_{j,k=0}^{n}\sum p_j \bar{s}_k c_{j-k} = (\pi, \mathbf{C}_n \sigma) , \qquad (2)$$

where we have denoted by $(\cdot, \cdot)$ the ordinary scalar product for vectors, by $\mathbf{C}_n$ the $(n+1) \times (n+1)$ Hermitian Toeplitz matrix whose k-th row has entries c_{j-k}, $0 \leq j \leq n$, and by π, σ the vectors $(p_0, \ldots, p_n)$ and $(s_0, \ldots, s_n)$ having as their components the coefficients of P and S, respectively; here and subsequently we write all vectors in row form. From (2), the vector π_n of coefficients of P_n is characterized by the equation

$$\mathbf{C}_n \pi_n = (0, \ldots, 0, \alpha)\,; \tag{3}$$

we normalize P_n so that $\alpha = 1$.

This association of a measure with moments and orthogonal polynomials gives rise to a rich variety of inverse problems, interesting in themselves and far-reaching in application. So for example it is well known [6, p. 43] that $P_n(z)$ is the n-th orthogonal polynomial for some $d\mu \geq 0$ if and only if all the zeros of P_n lie in $|z| < 1$, and that one measure which so generates $P_n(z)$ is

$$d\mu = |P_n(e^{i\theta})|^{-2}\, d\theta\,.$$

When $d\mu$ is a signed measure, the quadratic form (1) no longer defines a Hilbert space: $[P, P]$ may be negative. Nevertheless, if

$$d_n \equiv \det \mathbf{C}_n \neq 0\,,$$

P_n can be constructed to satisfy (3), and one can again ask for a characterization of such polynomials. In 1966, M. G. Krein [3] answered this question, also in terms of the zeros.

Theorem 1 (M. G. Krein). *A necessary and sufficient condition for $P_n(z)$ to be the n-th orthogonal polynomial corresponding to some measure $d\mu = f(\theta)d\theta$ is that the leading coefficient be real and that the set of zeros of P_n be disjoint from its reflection in $|z| = 1$.*

Theorem 2 (M G. Krein). *If $d_i \neq 0$, $0 \leq i \leq n-1$, let β_n, γ_n denote, respectively, the number of permanences and changes of sign in the sequence $1, d_0, \ldots, d_{n-1}$. Then P_n has β_n or γ_n zeros in $|z| < 1$, as $d_{n-1}\, d_n$ is positive or negative, respectively.*

Two elementary treatments of these results have recently appeared. The first [2] proves both theorems by exploiting properties of Toeplitz matrices; in Theorem 1, it gives an explicit formula for $\mathbf{C}_n^{-1}$ from the coefficients of P_n. The second [1] addresses Theorem 1 alone, relying on nothing but simple facts about Fourier series; it shows how to determine a measure which generates P_n, and extends the construction to polynomials having matrices as coefficients. While it is such generalizations that are at the forefront of current research [**4**], the scalar case retains some interest, in itself and as a possible springboard for new development. We therefore give here yet another proof of Theorems 1 and 2. Our approach resembles that of [**2**], but uses only rudimentary features of polynomials and of linear decomposition to generate a quadrature formula for (1) from $P_n(z)$, which in turn yields $\mathbf{C}_n$ directly.

Acknowledgment

It is a pleasure to thank Israel Gohberg for having introduced me to this problem, and for many interesting conversations.

A space of polynomials

Suppose that $\mathbf{C}_n$ is an $(n+1)\times(n+1)$ hermitian Toeplitz matrix with $d_n \equiv \det \mathbf{C}_n \neq 0$, and let (2) define an indefinite scalar product in $\mathcal{P}_n$, the linear space of polynomials of degree n. The Toeplitz nature of $\mathbf{C}_n$ can be succinctly expressed by

$$[zA, zB] = [A, B]. \tag{4}$$

if $[A, B] = 0$, we will call A and B *orthogonal*, despite the fact that this now has no geometric meaning since A can well be orthogonal to itself. Nevertheless, as in Hilbert space, the collection of elements orthogonal to a k-dimensional subspace Γ_k has dimension $n+1-k$, for if

$$0 = [S, A] \equiv (\sigma, \mathbf{C}_n \alpha)$$

for all $S \in \Gamma_k$, then $\{\mathbf{C}_n \alpha\}$ forms a subspace of dimension $(n+1-k)$, and since $\mathbf{C}_n$ is invertible, so does the subspace $\{\alpha\}$. This fact leaves unchanged all the familiar linear relationships which hold among polynomials orthogonal in a positive definite scalar product, and so the discussion can now follow that of [7], which we next sketch briefly for completeness.

As we have seen, if π_n satisfies

$$\mathbf{C}_n \pi_n = (0, \ldots, 0, 1)$$

then by (2) the polynomial $P_n(z)$ having the components of π_n as coefficients is orthogonal to all polynomials of lower degree, and, denoting by t_n the leading coefficient of P_n,

$$[P_n, P_n] = (\pi_n, \mathbf{C}_n \pi_n) = t_n = d_{n-1}/d_n, \tag{5}$$

the last by Cramer's rule.

Evaluations

By the symmetry of $\mathbf{C}_n$, if corresponding to $\pi_n = (a_0, \ldots, a_{n-1}, t_n)$ we introduce $\pi'_n = (\bar{t}_n, \bar{a}_{n-1}, \ldots, \bar{a}_0)$, we obtain

$$\mathbf{C}_n \pi'_n = (1, 0, \ldots, 0). \tag{6}$$

Thus on setting

$$E_n^0(z) \equiv z^n \bar{P}_n \left(\frac{1}{z}\right) \tag{7}$$

we get a polynomial corresponding to π'_n whose effect in the scalar product, by (2) and (6), is to evaluate the other member at zero:

$$[S_n, E_n^0] = \sigma_0 = S_n(0).$$

More generally, let the *evaluation* $E_n^\zeta(z)$ be the polynomial which similarly evaluates at $z = \zeta$:

$$[S_n, E_n^\zeta] = S_n(\zeta),$$

for each $S_n \in \mathcal{P}_n$. On writing $E_n^\zeta = a P_n + Q_{n-1}$, with Q_{n-1} of degree $n-1$, and forming the scalar product on both sides with P_n, we find, by definition of E_n^ζ and P_n,

$$P_n(\zeta) = [P_n, E_n^\zeta] = \bar{a}[P_n, P_n] + [P_n, Q_{n-1}] = \bar{a}\, t_n\,,$$

whence

$$a = \overline{P_n(\zeta)}/t_n\,. \tag{8}$$

Again by definition of E_n^ζ and by (4), for each $S_{n-1} \in \mathcal{P}_{n-1}$,

$$\begin{aligned} 0 = [(z-\zeta)S_{n-1}, E_n^\zeta] &= [z\,S_{n-1}, E_n^\zeta] - [S_{n-1}, \bar{\zeta} E_n^\zeta] \\ &= [z\,S_{n-1}, (1-\bar{\zeta} z)\,E_n^\zeta]\,. \end{aligned} \tag{9}$$

By (8), the combination $(1-\bar{\zeta}z)E_n^\zeta + \bar{\zeta}\, z\, \overline{P_n(\zeta)}\, P_n(z)/t_n$ lies in $\mathcal{P}_n$ since the leading terms cancel, and by (9), (4), and the definition of P_n, it is orthogonal to the n-dimensional subspace $\{z\, S_{n-1}\}$, as is E_n^0. Consequently,

$$(1-\bar{\zeta}\, z)E_n^\zeta(z) + \bar{\zeta}\, z\, \overline{P_n(\zeta)}\, P_n(z)/t_n = \gamma\, E_n^0(z)\,,$$

and on evaluating at $z = 1/\bar{\zeta}$ to determine γ, and simplifying by means of (7), we obtain

$$E_n^\zeta(z) = \frac{E_n^0(z)\,\overline{E_n^0(\zeta)} - z\, P_n(z)\, \bar{\zeta}\, \overline{P_n(\zeta)}}{t_n(1-z\,\bar{\zeta})}\,. \tag{10}$$

This is known as the Christoffel-Darboux formula. It can be written in matrix form [7] to give the formula of Gohberg-Semencul for $\mathbf{C}_n^{-1}$ [**5**, p. 86] which yields the expression in [**2**].

Just as $P_n(z)$ determines evaluations by means of (10), so does a set of $n+1$ evaluations determine $P_n(z)$. For with given points $\zeta_1, \ldots, \zeta_{n+1}$, set

$$R(z) \equiv (z-\zeta_1) \cdots (z-\zeta_{n+1})\,;$$

then by a contour integration, for any $S_{n-1} \in \mathcal{P}_{n-1}$,

$$\sum_{k=1}^{n+1} \frac{S_{n-1}(\zeta_k)}{R'(\zeta_k)} = \lim_{r\to\infty} \frac{1}{2\pi i} \int_{|z|=r} \frac{S_{n-1}(z)}{R(z)}\, dz = 0,$$

since the integrand is $O(r^{-2})$. This means that $\sum E_n^{\zeta_k}(z)/\overline{R'(\zeta_k)}$ is orthogonal to $\mathcal{P}_{n-1}$, hence is a constant multiple of P_n; by forming the scalar product with $R'(z) \in \mathcal{P}_n$ we see that the constant of proportionality is 1. Thus

$$P_n(z) = \sum \frac{E_n^{\zeta_k}(z)}{\overline{R'(\zeta_k)}}. \tag{11}$$

We remark that, for specially chosen sets $\{\zeta_k\}$ we can prove (11) by purely linear means, without the use of contour integration. Specifically, for a point ζ in the complex plane, denote by

$$\zeta^* = 1/\bar{\zeta}$$

its reflection in $|z| = 1$, suppose that the set of points $\{\zeta_k\}$ coincides with $\{\zeta_k^*\}$, and that $E^{\zeta_k}(\zeta_j)$ vanishes except when $\zeta_j = \zeta_k^*$. Such sets of evaluations always exist, being generated from (10) at points which are distinct zeros of $zP_n(z) - \rho E_n^0(z)$, with $|\rho| = 1$, and (11) follows from the fact that $(z-\zeta_k^*)E_n^{\zeta_k}$ coincides with $\overline{P_n(\zeta_k)}\,R(z)$.

Finally, if $d_0, d_1, \ldots, d_n$ are all nonzero, we can define the entire sequence of orthogonal polynomials $P_0(z), P_1(z), \ldots, P_n(z)$ and, correspondingly, $\{E_k^0(z)\}$, the evaluations at $z=0$ for polynomials of degree k. The polynomial $zP_k(z)/t_k - P_{k+1}(z)/t_{k+1}$ then lies in $\mathcal{P}_k$ since the leading terms cancel, and by definition of P_k and (4) it is orthogonal to the k-dimensional subspace of $\mathcal{P}_k$ consisting of $\{zS_{k-1}\}$, as is E_k^0. Consequently, these are proportional, so that

$$\frac{P_{k+1}(z)}{t_{k+1}} + \nu_k \frac{E_k^0(z)}{t_k} = \frac{zP_k(z)}{t_k} \tag{12}$$

for some constant ν_k. By forming the scalar product of each side with itself, using the orthogonality of P_{k+1} to $\mathcal{P}_k$, (4), and (5), we find

$$\frac{1}{t_{k+1}} + \frac{|\nu_k|^2}{t_k} = \frac{1}{t_k}, \tag{13}$$

whence $|\nu_k| \neq 1$.

Theorems 1 and 2

From these preliminaries, we now pass to a proof of Theorems 1 and 2. For Theorem 1, suppose first that α and α^* are zeros of $P_n(z)$; the eventuality $\alpha = \alpha^*$ is not excluded. Then

$$P_n(z) = (z-\alpha)\,Q_{n-1}(z) = (1-\bar{\alpha}z)\,T_{n-1}(z), \tag{14}$$

so that $|Q_{n-1}(z)| = |T_{n-1}(z)|$ on $|z| = 1$, and since, by (1), $[S, S]$ depends only on the values of $|S|$ on $|z| = 1$, we conclude that

$$[Q_{n-1}, Q_{n-1}] = [T_{n-1}, T_{n-1}]. \tag{15}$$

Now on rewriting (14) as

$$P_n(z) + \alpha\, Q_{n-1}(z) = z\, Q_{n-1}(z),$$

and forming the scalar product of each side with itself we find, by the orthogonality of P_n to Q_{n-1}, and by (4),

$$[P_n, P_n] + |\alpha|^2[Q_{n-1}, Q_{n-1}] = [z\,Q_{n-1}, z\,Q_{n-1}] = [Q_{n-1}, Q_{n-1}], \tag{16a}$$

and similarly

$$[P_n, P_n] + [T_{n-1}, T_{n-1}] = |\alpha|^2[T_{n-1}, T_{n-1}]. \tag{16b}$$

On adding (16a) and (16b) and applying (15),

$$2[P_n, P_n] = (|\alpha|^2)\Big([Q_{n-1}, Q_{n-1}] - [T_{n-1}, T_{n-1}]\Big) = 0,$$

a contradiction.

For the converse, suppose that $Q(z)$, of degree n, has no zeros which are reflections of each other in $|z| = 1$, and is normalized so its leading coefficient q_n is real. Thereupon $F(z)$, defined analogously to (7) by

$$F(z) \equiv z^n \, \overline{Q} \left(\frac{1}{z} \right), \tag{17}$$

has no zeros in common with $Q(z)$. Since $F(0) = q_n \neq 0$, $F(z)$ likewise has no zeros in common with $z\,Q(z)$. Now choose ρ with $|\rho| = 1$ so that the equation

$$\frac{z\,Q(z)}{F(z)} = \rho$$

has $n+1$ distinct roots $\xi_1, \ldots, \xi_{n+1}$; by (17) if ξ is a root, then so is ξ^*, since $\rho = \rho^*$. Motivated by the behavior of (10), we set

$$R(z) = \prod_{k=1}^{n+1} (z - \xi_k) = q_n^{-1} [zQ(z) - \rho F(z)],$$

and introduce

$$F^{\xi_j}(z) \equiv \overline{Q(\xi_j)} \, \frac{R(z)}{z - \xi_j^*}, \qquad 1 \leq j \leq n+1. \tag{18}$$

The coefficient $Q(\xi_j)$ does not vanish, else, since $R(\xi_j) = 0$, we have also $F(\xi_j) = 0$, contradicting the assumption that Q and F have no zeros in common. We note first that

$$\overline{F^{\xi_j}(\xi_j^*)} = F^{\xi_j^*}(\xi_j), \tag{19}$$

that is, by (18), that

$$\overline{R'(\xi_j^*)} \,/\, R'(\xi_j) = \overline{Q(\xi_j^*)} \,/\, Q(\xi_j). \tag{20}$$

We can prove (20) without calculation by writing $q_n R'(\xi_k) \,/\, Q(\xi_k)$ as the value at $z = \xi_k$ of $d(\log zQ/F) \,/\, d \log z$, and applying the reflection principle to this function, which is real on

$|z| = 1$ by (17). But also, directly, by the identity $\overline{(\lambda^* - \mu)} = -\bar{\mu}(\lambda - \mu^*)/\lambda$,

$$\overline{R'(\xi_j^*)} = \prod_{\xi_k \neq \xi_j^*} \overline{(\xi_j^* - \xi_k)} = \frac{(-1)^n R'(\xi_j)}{(\xi_j)^n} \prod_{\xi_k \neq \xi_j^*} \bar{\xi}_k,$$

while on evaluating R at $z=0$ we find $(-1)^{n+1} \prod \xi_k = -\rho$, so that

$$\overline{R'(\xi_j^*)} / R'(\xi_j) = \bar{\rho}\xi_j / (\xi_j)^n = \bar{\rho} / (\xi_j)^{n-1}.$$

Simultaneously, from (17), $\overline{Q(\xi_j^*)} = F(\xi_j) / (\xi_j)^n$, whence

$$\frac{\overline{Q(\xi_j^*)}}{Q(\xi_j)} = \frac{F(\xi_j)}{\xi_j Q(\xi_j)} \frac{1}{(\xi_j)^{n-1}} = \frac{1}{\rho(\xi_j)^{n-1}};$$

since $|\rho| = 1$, this establishes (20). Moreover, by (18), we see that $\{F^{\xi_j}\}$ form an interpolating set at the points $\{\xi_j^*\}$ and so

$$Q(z) = \sum_{k=1}^{n+1} Q(\xi_k^*) \frac{F^{\xi_k}(z)}{F^{\xi_k}(\xi_k^*)} = \sum \frac{F^{\xi_k}(z)}{\overline{R'(\xi_k)}}, \tag{21}$$

the last equality by writing (18) for ξ_k^*, letting $z \to \xi_k$, and applying (19).

The crux of the argument now is to define an indefinite scalar product in $\mathcal{P}_n$ by

$$[S, T] = \sum_{1}^{n+1} \frac{S(\xi_k^*)\overline{T(\xi_k)}}{F^{\xi_k}(\xi_k^*)}; \tag{22}$$

by (19), and because the sets $\{\xi_k\}$ and $\{\xi_k^*\}$ coincide, this is Hermitian, as required. The matrix which generates this scalar product from coefficients of the polynomials S and T has entries

$$c_{p,q} = [z^p, z^q] = \sum \frac{(\xi_k^*)^p(\bar{\xi}_k)^q}{F^{\xi_k}(\xi_k^*)} = \sum \frac{(\xi_k^*)^{p-q}}{F^{\xi_k}(\xi_k^*)}, \tag{23}$$

hence is a Toeplitz matrix. From (22), the interpolating property of $F^{\xi_j}(z)$, and (19), the evaluation $E_n^{\xi_j}(z)$ evidently coincides with $F^{\xi_j}(z)$, since, by (22), $[S, F^{\xi_j}]$ reduces to the single term of the sum corresponding to $\xi_k = \xi_j^*$, so that

$$[S, F^{\xi_j}] = \frac{S(\xi_j)\overline{F^{\xi_j}(\xi_j^*)}}{F^{\xi_k^*}(\xi_k)} = S(\xi_j).$$

Now by (11), the n-th orthogonal polynomial here is

$$P_n = \sum \frac{F^{\xi_k}(z)}{\overline{R'(\xi_k)}},$$

and we see from (21) that $P_n(z) = Q(z)$. We have thus exhibited the given $Q(z)$ as the n-th orthogonal polynomial corresponding to a Toeplitz matrix $\mathbf{C}_n$, whose entries are given explicitly by (23). This matrix is nonsingular, since the evaluations F^{ξ_k}, $1 \le k \le n+1$, are linearly independent. A corresponding real measure which generates the $\{c_k\}$ as moments is $d\mu(\theta) = \sum_{|k| \le n} c_k e^{ik\theta} d\theta$. This establishes Theorem 1. Since $\bar{z} = 1/z$ on $|z| = 1$, the integrand of (1) can be written as $S_n(z)\bar{T}_n(1/z)$, that is, as a trigonometric polynomial of degree n,

$$H_n(z) = \sum_{k=-n}^{n} Q_k z^k,$$

and, conversely, every such trigonometric polynomial can be factored as $S_n(z)\bar{T}_n(1/z)$, for some choice of S and T. Now for the terms in (22) we have

$$S(\xi_k^*)\overline{T(\xi_k)} = S(\xi_k^*)\bar{T}(\bar{\xi}_k) = S(\xi_k^*)\bar{T}(1/\xi_k^*) = H_n(\xi_k^*).$$

Thus (22) is a *quadrature formula*

$$\int_{|z|=1} H_n(z)d\mu = \sum_{k=1}^{n+1} H_n(\xi_k^*)/E_n^{\xi_k}(\xi_k^*),$$

for integrating trigonometric polynomials on $|z| = 1$ with respect to a real signed measure.

To prove Theorem 2, if δ_n is the number of zeros of P_n in $|z| < 1$, then, since E_n^0 reflects them in $|z| = 1$, E_n^0 has $(n - \delta_n)$ zeros in $|z| < 1$. From (12),

$$\frac{t_k}{t_{k+1}} \, \frac{P_{k+1}(z)}{z\,P_k(z)} = 1 - \nu_k \, \frac{E_k^0(z)}{z\,P_k(z)} \,,$$

with $|\nu_k| \neq 1$. Since, by (7), $|E_k^0 / z\,P_k| = 1$ on $|z| = 1$, the argument principle gives

$$\delta_{k+1} - (1+\delta_k) = \Delta_{|z|=1} \arg \frac{P_{k+1}}{z\,P_k} = \begin{cases} 0\,, & |\nu_k| < 1\,, \\ (k-\delta_k) - (1+\delta_k)\,, & |\nu_k| > 1\,. \end{cases} \tag{24}$$

We can reach the same conclusion by applying (7) to (12), obtaining

$$\frac{E_{k+1}^0(z)}{t_{k+1}} = \frac{E_k(z)}{t_k} - \bar{\nu}_k \, \frac{z\,P_k(z)}{t_k} \,, \tag{25}$$

setting $B_k = P_k / E_k^0$, rewriting (12) to parallel (25) and dividing the two equations to see that the Schur algorithm generates the sequence $\{B_k\}$. By (13), $|\nu_k| < 1$ if and only if $t_k / t_{k+1} > 0$. By (5), $d_k = d_{k-1} / t_k$, so $\operatorname{sign}(d_k d_{k-1}) = \operatorname{sign} t_k$; thus the number of permanences, and changes, of sign in the sequence $1, d_0, \dots, d_{n-1}$ is the number of positive and negative t_j in the sequence $t_0, \dots, t_{n-1}$. To count these, define

$$\eta_k = \begin{cases} \text{no. of } t_j > 0 \text{ among } t_0, t_1, \dots, t_{k-1}\,, & \text{if } t_k > 0\,; \\ \text{no. of } t_j < 0 \text{ among } t_0, t_1, \dots, t_{k-1}\,, & \text{if } t_k < 0\,. \end{cases}$$

If t_{k+1} and t_k have the same sign, then clearly $\eta_{k+1} = 1 + \eta_k$. If $t_{k+1} > 0$ while $t_k < 0$, then the number of positive t_j among $t_0, \dots, t_k$ is the same as that among $t_0, \dots, t_{k-1} = k - \eta_k$, and similarly if $t_{k+1} < 0$ with $t_k > 0$. Thus

$$\eta_{k+1} = \begin{cases} 1 + \eta_k\,, & t_k / t_{k+1} > 0\,; \\ k - \eta_k\,, & t_k / t_{k+1} < 0\,. \end{cases}$$

This is precisely (24), and so $\delta_k = \eta_k$, establishing Theorem 2.

REFERENCES

[1] Atzmon, A., *N-orthonormal operator polynomials*, This volume.

[2] Ellis, R. L., I. Gohberg, and D. C. Lay, *On two theorems of M. G. Krein concerning polynomials orthogonal on the unit circle*, Integral Eq. Oper. Theory, 1 (1988) 87-104.

[3] Krein, M. G., *On the distribution of roots of polynomials which are orthogonal on the unit circle with respect to an alternating weight*, Teor. Funkcii Funkc. Analiz. i Prilozen. Resp. Sb., No. 2 (1966), 131-137 (Russian).

[4] Gohberg, I. and L. Lerer, *Matrix generalizations of M. G. Krein's theorems on orthogonal polynomials*, This volume.

[5] Gohberg, I. and D. A. Fel'dman, *Convolution equations and projection methods for their solution*, Transl. Math. Monographs, Vol. 41, Amer. Math. Soc., Providence, RI, 1974.

[6] Grenander, U. and G. Szegö, *Toeplitz forms and their applications*, Univ. of Calif. Press, Berkeley, CA, 1957.

[7] Landau, H. J., *Maximum entropy and the moment problem*, Bull. Amer. Math. Soc., **16** (1987), 47-77.

AT&T Bell Laboratories
Murray Hill, New Jersey 07974